OTHER BOOKS BY DESMOND MORRIS

The Biology of Art

Men and Snakes (*co-author*)

Men and Apes (*co-author*)

Men and Pandas (*co-author*)

The Mammals: A Guide to the Living Species

Primate Ethology (*editor*)

The Naked Ape

The Human Zoo

Patterns of Reproductive Behaviour

Intimate Behaviour

Manwatching

Gestures (*co-author*)

Animal Days

The Soccer Tribe

The Book of Ages

The Art of Ancient Cyprus

BODYWATCHING

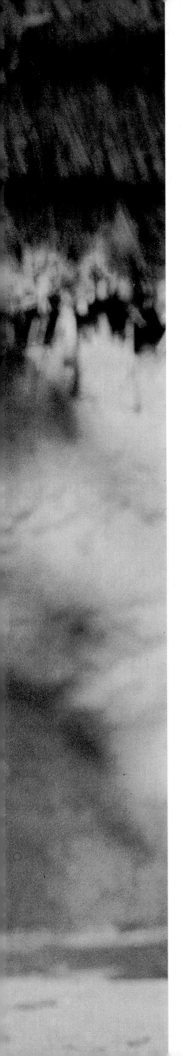

DESMOND MORRIS

BODYWATCHING
A Field Guide to the Human Species

JONATHAN CAPE THIRTY-TWO BEDFORD SQUARE LONDON

First published 1985

Jonathan Cape Ltd
32 Bedford Square
London WC1B 3EL

Text copyright © 1985 by
Desmond Morris
Compilation copyright © 1985
by Equinox (Oxford) Ltd

AN EQUINOX BOOK
Planned and produced by
Equinox (Oxford) Ltd, Littlegate
House, St Ebbe's Street, Oxford,
OX1 1SQ, England, in cooperation
with Jonathan Cape Ltd

British Library Cataloguing
in Publication Data
Morris, Desmond.
    Bodywatching: a field guide
to the human species.
    1. Psychology
I. Title
150   BF121

ISBN 0–224–02844–8

Text films set by Peter MacDonald,
Hampton, Middlesex

Origination by Alpha
Reprographics Ltd, Harefield,
Middlesex; and Novacolour Ltd,
Digbeth, Birmingham

Printed in Holland by Royal
Smeets Offset, Weert

THE CONTENTS

BODYWATCHING

Nothing fascinates us quite as much as the human body. Whether we realize it or not, we are all obsessed with physical appearances. Even when we are engaged in lively conversation and seem to be engrossed in purely verbal communication we remain ardent bodywatchers.

By the time we are adults we are all highly sensitive to the tiniest changes in expression, gesture, posture and bodily adornment of our companions. We acquire this sensitivity in a rough-and-ready fashion, through intuition rather than analysis. If we took the trouble to make a more analytical study of body appearances we could become even more sensitized to them, and could avoid some of the pitfalls into which our intuition sometimes leads us.

Clever as we are at body-reading, we do make errors. Often they are serious errors which amount to superstitions about our bodies. These become so entrenched that they are hard to eradicate from popular thought. As we shall see, it is surprising how many of our cherished ideas about parts of the human body are based on false premises.

Also blurring our bodywatcher's vision is the fact that we take certain aspects of our own bodies very much for granted. We are so familiar with the figure we see reflected in the bathroom mirror that we cease to ask questions about the way the human animal came to be the strange shape it is. We may become worried about our weight or our health and start to fret about diets and exercise, but that is another matter. Such anxiety-ridden preoccupations do not help us to see ourselves in an evolutionary perspective. They stop short of pinpointing us as part of a wider zoological scene.

When a natural history programme on television shows

■ The human body is one of the strangest objects in the natural world, strutting about on its hind legs, functionally naked, with a soft fleshy surface easily damaged by a sharp thorn, with no claws or fangs, no armoured scales and no poison glands – and yet it has subjugated the whole planet. All the human bodies on Earth at present belong to one and the same species Homo sapiens, although in the past there have been other, closely related species which have fallen by the wayside (or perhaps been thrown there). Today, the Caucasian and Oriental races make up 92 per cent of the world population, with other groups, such as Negroes and Bushmen, accounting for the other 8 per cent.

■ Some authorities have over-stressed the racial differences between groups of people. Even the more conspicuous of these differences are comparatively minor in terms of human biology, having little to do with anything except adaptations to different degrees of sunlight, temperature and humidity. With the advent of clothing and architecture, central heating and air conditioning, even these small differences have lost their significance. All human bodies are remarkably similar and so, for that matter, are human personalities. The variations in both are fascinating, of course, but they are slight in comparison with the similarities, as anyone who travels around the world with an objective, un-biased eye soon discovers.

■ The human life cycle is a drama in three acts. The body starts as a tiny egg about the size of one of these dots.... This egg is itself 2,000 times bigger than the sperm that fertilizes it to produce a single cell called a zygote. This starts to grow and divide until after about 266 days it emerges from the womb, wriggling and gasping and ready for Act II. This sees the infant grow to physical maturity in 25 years (by which time the original cell will have multiplied literally millions of millions of times) and then into mental maturity in another 10-15 years. In the interval before the final act, there is usually a brief midlife crisis (around the age of 40) as the body discovers that the best has already been and has somehow slipped by so quickly. Act III sees a steady decline from middle age to senility and death in about another 30-40 years, if nothing untoward happens to interrupt the natural course of events. The decline is not as bad as it sounds if the body is still in reasonable shape – just a gentle mellowing with its own special pleasures and magic moments, less fiery and fierce, but with plenty to savour.

At birth the body is between 19 and 22 inches long and weighs between six and eight pounds. The adult weight will be roughly twenty times greater. By the age of five, the infant is 43 inches tall and weighs about 40 pounds. It can now speak 2,000 words and is about to plunge into a long period of childhood schooling.

By the age of 18 both sexes will have reached their adult height and weight. The average figures are: for the male: 5 feet 9 inches and 162 pounds; and for the female: 5 feet 4 inches and 135 pounds.

us an unfamiliar exotic species such as a spotted sea-hare or a checkered elephant-shrew, we marvel at its bizarre structure and its peculiar colouring, its intricate design and its complex movements, but it seldom strikes us that the human body itself is the most unusual and intriguing organism in the whole animal kingdom.

If intelligent aliens had visited this planet a few million years ago and looked at the monkeys scampering about in the tree-tops, they could hardly have guessed that in a twinkling of evolution's eye the descendants of one of these chattering monkeys would be playing the piano with their front toes and planting a flag on the moon. The amazing human success story would have been impossible to predict.

It is still difficult to determine precisely how we came to abandon the furry, four-footed way of life and replace it with a naked-skinned, bipedal existence. Why *did* we rear up and start strutting about on our hind legs? It is a top-heavy, ungainly posture which allows us only half the speed of any fleet-footed monkey. Yet with it we have managed to subjugate the whole planet. How did it begin?

Some authorities see it as an adaptation to a wading, swimming, aquatic way of life, with the 'aquatic ape' returning eventually to dry land equipped with a layer of blubber beneath its streamlined skin surface, and with a shellfish hunter's taste for meat. The more orthodox view prefers a direct shift from forest fruit-picking to chasing prey on the plains. One scenario pictures this hunting ape adopting a vertical posture to scan the distant horizons over the tops of the tall grasses in its new habitat. Another suggests that the vital change came about as a result of the need to carry larger food objects. Chimpanzees faced with large bunches of bananas rear up on their hind legs to carry them off. A third idea is that it was the need to hold and throw prey-killing weapons that demanded a vertical posture.

Whatever the pressures on our distant forebears, the fact

By the age of 25 the final physical growth patterns are completed, these being concerned with the fusion and strengthening of the bones. This is the age of peak muscle condition – the age for athletes and sportsmen. The mental powers continue to grow and develop, however, reaching a peak of inventiveness and creativity in the period between 35 and 40 – although in certain special spheres, such as music and mathematics, genius flowers much earlier.

By the age of 50 the level of the male sex hormones is beginning to fall off and the frequency of sexual activity is down to about half its youthful figure. Females are starting to experience the menopause – the average age for this cessation of ovulation being 51.7 years.

Sixty-nine is the average lifespan of the English-speaking male. The female lives longer – to 75 in Britain and to 77 in the United States. Those men and women who do manage to live into their eighties will find that their bones are increasingly brittle and susceptible to breaks, that their skin will lose its elasticity, and that in many cases their short-term memory fails them. But great achievements have been made by humans still in their eighties and nineties – and even past 100. The longest lifespan officially recognized at present is that of a Mr Izumi in Japan, who celebrated his 120th birthday in June 1985.

remains that they did become bipedal despite a loss of running speed, and this altered their bodies in many ways. The hidden underside of the hairy monkey became the full-frontal display of the naked human. The back became the rear. The pelvic and neck angles changed dramatically. The front feet became grasping hands which not only carried but also manipulated and gestured. A unique body shape was evolving, strikingly different from anything that had gone before it and devastatingly effective in its impact on the environment.

Although the upright posture made it less easy to give birth, the strange new human species was soon breeding at a startling rate. Tribes grew and split and spread over the whole face of the globe. The front feet that had become busy hands fashioned tools and weapons, buildings and vehicles, and this peculiar 'walking ape' soon came to dominate its environment. Ten thousand years ago there were already as many as ten million human beings dispersed over almost all the land surfaces of the planet. Today even that huge number seems insignificant, the population of London alone being that great. By the year 2000, which is already so close, it is estimated that there will be 6,082 million human bodies crowding the earth. Despite the way this incredible success story has raced ahead the human body at the centre of it has remained much the same. If a prehistoric baby from, say, 40,000 years ago were to be whisked into the present day by a

■ *The human body comprises a skeleton of 208 bones, with a total weight of about 20 pounds; more than 600 muscles, which make up roughly 35-45 per cent of the total body weight; a blood system containing between 9 and 12 pints of blood, operated by a heart which during a lifetime does enough work to have lifted a ton weight 150 miles up into the air; a nervous system dominated by a brain which makes the biggest computer look like a child's toy – and which programmes itself; a pair of lungs which handle 500 cubic feet of air a day; a cooling system involving between two and three million sweat glands; a feeding system which boasts a 25-foot-long alimentary canal and which deals with the conversion of 50 tons of food in an average lifetime; a reproductive system that has all too successfully populated today's world with more than 4,000 million human beings; an excretory system with kidneys capable of filtering 45 gallons of fluid a day; and 17 square feet of skin to cover everything and, as one doctor put it, 'to keep the blood in and the rain out.'*

This is the pulsating, complex and yet vulnerable organism that we have taken to the depths of the ocean and up to the moon. It is the animal which has invented language, art, science, sport, architecture, politics and religion. It has conquered the world and may yet destroy it through a surfeit of success.

■ Although all human bodies are basically the same, differences in body height are particularly noticeable. There have been some amazing variations in size. The tallest man ever recorded was an American giant who reached the staggering height of 8 feet 11 inches and the tallest woman, who was from China, was 8 feet 1 inch. The shortest woman, who was from Holland, was a tiny 23·2 inches. This means that the tallest human beings have been more than four times the height of the shortest.

With body weight the contrast is even more extraordinary. The heaviest human being was a vast American who, it was calculated, topped 1,400 pounds. He was too huge to weigh accurately. The lightest was a painfully thin female midget from Mexico who, at 17, weighed no more than 4·7 pounds. This means that the bulkiest human being was 300 times heavier than the lightest. It is difficult to conceive of such people belonging to the same species, but in all respects except size there will have been remarkably little difference between them.

time machine and reared by a modern family, nobody would notice the difference.

Our basic behaviour has also altered remarkably little with the passage of time. Although over the centuries there has been a great deal of preaching and theorizing by priests, politicians and professors about the way human beings ought to conduct themselves, their earnest attempts to pull or push us in this direction or that have left no deep impression. Some people find this hard to believe, insisting that the savage brutes of yesterday have become the civil citizens of today, or conversely that the innocently noble savages of yesterday have become the warped sadistic thugs of today. Such simplifications are meaningless, as are the claims of generations of moralists that their teaching has improved us and made us good.

The truth is that the human species has always possessed the same set of emotional urges and basically the same way of expressing them externally. We have always been capable of swinging from hostility to friendliness, from love to hate, from selfishness to altruism, from sadness to joy. All that has happened is that the names have been changed along the way: for hunting urge read work ethic; for pecking order read class struggle; for outbreeding read incest taboo; for pair bond read marriage; for tribal identity read cultural heritage; and so on. We like to think that a laudable quality such as mutual aid, for example, represents a new form of civilized enlightenment, but it is as old as the primeval hunt, which forced us to cooperate or die. In the past I have been accused of saying that human beings are full of 'beastly' feelings, but this criticism is based on muddled thinking. In reality most of our 'finer feelings' *are* beastly ones in the sense that they are part of our animal inheritance. We do not need religious rule books or codes of ethics to make us caring, loving individuals. It is in our animal nature.

Of course, we are competitive and querulous, as well. That too is part of our animal personality and the balance is a fine one, easily tipped over into violence and bloodshed by artificially stressful surroundings. When this happens our human bodies suddenly seem painfully frail, threatened on all sides by knives and guns and bombs, flying glass and falling concrete. We constantly struggle to control these explosions of violence and to maintain a home environment in which the vulnerable, fleshy, sweating human body can survive in comfort. Hopelessly overcrowded as we are, it is not easy. We make many mistakes and many human bodies are damaged or destroyed in the process.

Some of our disasters could be avoided if the leaders of modern societies understood a little better the kind of spe-

cies to which they and their followers belong. Unfortunately they study far too little the very beings – the human beings – whom they claim to serve. If one were to design the ideal citizen, as seen by any given ruling faction, he or she would almost certainly be a far cry from the real thing. For example, some cultures have failed to recognize the human need for personal territory; others have underestimated the urge to form loving family units; still others have ignored the constructively rebellious nature of human curiosity and creativity. Sooner or later such errors cause social disquiet and the leaders in question suffer the consequences. At the heart of their ignorance lies a lack of knowledge about the way the human body itself operates as a social being. This is not a medical problem, concerned with internal organs or health. It is a matter of how we use our bodies when we encounter one another in everyday life. For this use reflects our innermost needs and desires – signalling these feelings to our companions and, if we look closely enough, telling us a great deal that we need to know about the true nature of mankind.

To ignore such knowledge today, when we have learned so much about human behaviour, is wanton. The statement that 'knowledge is sinful' has always been a desperate lie. Ignorance has always led to agony – to cruel superstitions, to needless anxiety, to religious bigotry and to mental repression. At the centre of all this, the human body has suffered endless unnecessary insults. Some of these have been wilfully imposed, others have been thoughtlessly self-inflicted. Part of the problem, as I said earlier, is that familiarity has bred complacency. Because we each own a human body we think we know all we *need* to know about it.

■ *During the course of our long evolution there was an increasing division of labour, with males specializing in hunting and females in food-gathering and child-rearing. The male and female bodies became altered slightly as adaptations to these more and more distinct roles. The males became more muscular and the females showed greater development of fat layers. Male respiration improved and the chest region expanded to house the larger lungs. Female hips widened in connexion with giving birth. Because these and many other sex differences are biological and basic there has been a tendency to amplify them by cultural means – making females super-feminine and male super-masculine. This reached a peak in earlier centuries, where the demands of breeding were still very heavy for nearly all women, but now, in an over-crowded world, we are starting to see the appearance of some far less feminine females than would have been socially acceptable in the past. With the burden of incessant pregnancy and childrearing removed, many urban women are adopting more and more independent pseudo-masculine roles, on the athletics field and even in the boxing ring.*

What is required to lift this veil of familiarity and enable us to examine the human body in a new light is some analytical device. The one I have chosen in this book is to treat the body surface as if it were a strange landscape and to explore it bit by bit, like a tourist investigating an exotic island. By isolating each part of the body and scrutinizing it in close-up it should be possible to regain a sense of wonder at the extraordinary animal called man. When this happens we should be less liable to carelessly maltreat ourselves or allow others to abuse our bodies.

Scanning the body from top to toe, the chapters which follow take the reader on a voyage of discovery – self-discovery – and reveal the complexity of the human being both as a biological entity and as a cultural phenomenon. The journey starts by examining the hair on top of the head, then moves on to the brow, the eyes, the ears, the nose and so on, down to the legs and the feet. Altogether there are twenty stops on the way and at each stop there is a chapter which examines the body-part in question from many different angles. First its anatomy, physiology, function, evolution and growth are briefly considered; then its behavioural possibilities are explored – its movements, postures, expressions and gestures; its most intriguing cultural modifications and exaggerations are examined – its adornment by painting and tattooing, its mutilation by shaving, cutting, piercing or scarring; and

■ *Like most primates, the human species has reduced its 'litter-size' to one, an evolutionary trend which sees more and more attention given to fewer and fewer offspring. As a result, a human baby has about the best chance of survival of any newborn animal. The price paid for this advantage is an enormous parental burden – the greatest of any living species. As with many kinds of birds and mammals where the parental duties are too heavy for the female alone, the human species has developed a pair-bond (the powerful sexual attachment we usually refer to as love) which activates paternal behaviour on the part of the breeding male.*

Because human offspring take so long to mature, the typical family unit has what could be described as a 'serial litter', with children of different ages overlapping one another in their demands for parental attention. This means that when the 'single-birth rule' of the human species is broken and a multiple birth occurs, the parents are put under severe strain. Twins are acceptable, even under primitive conditions, because the human female has two breasts, but beyond twins there are special problems. The chances of this happening are, however, rare. Worldwide, a mother has about a one-in-a-hundred likelihood of producing twins. For triplets it is about one in 10,000 and for quadruplets about one in a million. Quintuplets are so rare (about two sets a year in the world) that it is difficult to give figures, but it is roughly one chance in 40 million. These figures vary from country to country quite considerably and can only be taken as a rough guide. And there appear to be some racial differences, with blacks having far more twins than whites.

The identity confusions caused when identical twins are born remind us of the value of the small individual differences in appearance which the rest of us possess. If we ever reached the stage of cloning individuals – creating whole groups of people who looked identical – the chaos it would cause in ordinary social interactions would be enormous.

■ At present there are more than 4,000 million people alive on the face of this tiring planet and it has been estimated that in another generation, by the year 2020, there will be a staggering 7,600 million. In terms of bodily appearance this creates two conflicting problems. Individuals living in huge towns and cities teeming with massed humanity feel an increasing urge to impose their own personal identity on the social environment in which they live. To keep their individuality alive in the swelling crowd they often amplify their inherited uniqueness of facial features, and other body details, with artificial additions. This is done with personal variations in clothing and coiffure – everything from bright neckties and T-shirts to special hairstyles, make-up or moustache trims. In addition there are many kinds of personal ornaments, such as earrings, finger-rings, necklaces, bracelets and decorative jewels and badges, which can increase the differences in visual apperance between the increasingly overcrowded urbanites. Against this trend, however, is the primeval urge to display a body that shows similarities with the other members of one's own 'tribal' group. In extreme cases this creates uniformity of an exaggerated type, where it is almost impossible to recognize individuals – as with certain military groups or sports teams. But for the majority it is possible to decorate the body in such a way that both ends are served, some decorations signalling allegiance to a group and others providing 'badges' of individuality.

sometimes its role in the superstitious world of magic and religious symbolism is also investigated.

This artificial method of separating the human body into discrete body-units, like zones on a human map, not only offers a refreshingly unfamiliar approach to a familiar subject, but also helps to 'de-complicate' the body. Because the body works together as a whole, its body language is complex and difficult to assess all at once. Take the face, for example. There are many subtle changes in facial expression – at first sight too many to analyse, but if each element of the face is studied separately the task becomes much easier. Later, like an engineer who has dismantled an engine and then put it together again, we will know much more about how it works than we did before we examined each dismantled part in turn.

A reassuring message emerges from this analytical approach to bodywatching, namely that, despite all our frenzied attempts to make our bodies look so different from one another, with our various cultural badges and body customs, we are all basically the same. Beneath the invented baubles and bangles, hairstyles and cosmetics, we share far more than we usually admit. We like to imagine that 'our tribe', whether it be in New Guinea or New York, is somehow better than and far removed from all others, but this is merely a tribal fantasy. When it comes to important matters – we are all one.

THE HAIR

■ The hair that grows on top of our heads is one of the strangest features of the human body. Imagine what it must have been like for our ancient ancestors before there were brushes and combs, scissors and knives, hats and clothes. For more than a million years we were running around with almost naked bodies topped by a great mass of overgrown fur. While the hair on our trunks and limbs shrivelled to insignificance and exposed the whole of that skin surface to the open air, the hair on our scalps sprouted into a huge woolly bush or a long swishing cape. Unadorned and unstyled we must have looked amazing to other primates. What manner of ape was this?

Inherent in this question is the solution to the puzzle as to why we have such unusual scalp hair. The answer is that it made us look very different from other primate species. It was our 'species signal', visible from afar. Our great bushy heads on top of our smooth naked bodies identified us immediately as human. Our extragavant tresses were carried like a flag.

It is easy to forget this today because our scalp hair has been borrowed by fashion as an extra gender signal. In almost every culture, males and females dress their hair in masculine and feminine styles. So all-pervasive is this trend that we can be forgiven for overlooking the fact that, before the onset of balding in males, the structure of male and female scalp hair is identical. We do, of course, have hairy gender signals – moustaches, beards, hairy chests and the rest; but on top of the head, where our hair growth is at its most luxuriant, we enjoy complete sexual equality, both as children and as young adults. A flamboyant head of hair is neither feminine nor masculine, it is *human* and it set us apart from our primate relatives as we evolved into a distinct species.

If this seems improbable, it is only necessary to look at the hair patterns of various monkeys and apes. Closely related species often show dramatic differences in the colour, shape and length of their head hairs. Some have coloured 'caps' of hair, strikingly differentiated from the rest of the head hair; others display long moustaches; still others sport impressive beards; some even are bald. Clearly, the tendency to employ head hair differences as specific 'labels' is commonplace among primates, so it should not be too surprising that our own species employed a similar identification mechanism. What *is* surprising is that we carried it so far. Our scalp hair display became so exaggerated that, as soon as we were sufficiently advanced, technologically, to attack it with knives and scissors we set about curtailing it in a thousand ways. We snipped it and cut it, shaved it and tied it, braided it and twisted it, and stuffed it away under hats and caps. It was as though we could not stand the encumbrance of our primeval hair pattern any longer. The great weight of our sprouting locks had to be lightened in some way.

It would be wrong to conclude from this that we became anti-hair. What happened was that we made a cunning exchange of old hair for new. In our primeval condition, the human scalp pattern relied on sheer bulk to label our species. This made us different, highly conspicuously different, so that it was an efficient label, but it also happened to be rather cumbersome. When we advanced to the stage of being able to style our hair, we were able to find many new ways of making it conspicuous, with strange shapes, colours and adorn-

■ *If it is left untrimmed, the head hair of human beings is the longest and most luxuriant of any primate species. This applies to both sexes, and differences in hair length between men and women are due solely to cultural interference. Contrary to popular opinion, there are no detectable anatomical differences between individual male and female hairs. Each individual hair on top of the head grows for roughly six years, reaching an untrimmed length of about 40 inches. It then enters a resting phase for several months, with no further growth, until eventually it falls out and is replaced by a new active hair. The hair grows from a small papilla at the base of a skin-pocket called a follicle. Shown here: a resting hair, with the papilla connexion greatly weakened prior to the hair dropping out; and an actively growing hair.*

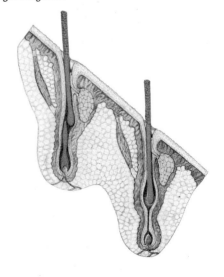

■ *Throughout history both men and women have cut, trimmed, shaped and styled their hair, nearly always making it shorter than its natural length. But our fascination with long flowing locks or great spreading bushes of luxuriant head hair remains strong, as though we retain some kind of deep-seated primeval response to the ancient human hair condition. Although hopelessly impractical for everyday use in modern times, exaggerated hair displays re-surface repeatedly among the more flamboyant young and in the studios of art photographers.*

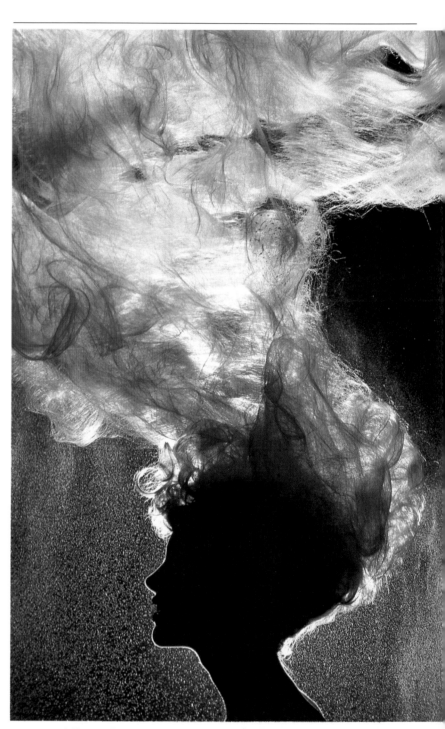

ments, while at the same time we could reduce its bulk. In modern times, with the whole paraphernalia of hats and wigs and hair-dressing, we could have the best of all worlds, having functional, working heads one moment and then fanciful display heads the next.

Before examining these new trends, it is important to take a closer look at the raw material on which they are based – the natural hair itself. On average, each human scalp boasts about 100,000 hairs. For some unknown reason, fair-haired people have more individual hairs than darker people. If you are blond you have about 140,000 hairs; if you are brown-haired, 108,000; but if you happen to be a

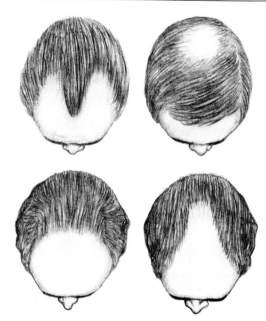

▲ *There are four main pathways to advanced baldness. They can be labelled as The Widow's Peak, The Monk's Patch, The Domed Forehead and The Naked Crown. In The Widow's Peak the hairline recedes more and more from the two temple regions, leaving a narrowing strip of hair along the centre-line of the head. In The Monk's Patch the frontal hairline holds firm but a bald spot starts to grow at the back of the top of the head. The baldness spreads steadily from that region. In The Domed Forehead the whole of the frontal hairline starts to recede, creeping farther and farther back, like the tide going out. In The Naked Crown the frontal hairline recedes fast in the central region and slower at the sides – the opposite of the Widow's Peak pattern. To complicate matters, these four main types of balding may be combined, so one individual may display two patterns at once. Genetic factors control these differences, and if a balding man looks at old photographs or paintings of his male ancestors he will usually find that his hair papillae are following in a long family tradition.*

redhead you are unlikely to have more than a mere 90,000.

Each hair grows from a small skin pocket called a follicle which has at its base a papilla. This minute lump of tissue is the hair-maker. It is rich in blood vessels and supplies the raw materials which are converted into the hair-cells. These cells keep forming on the surface of the papilla, the new ones pushing up the old ones, making the hair grow in length. Eventually the hair-root (the part beneath the skin surface) becomes so long that the tip of the hair emerges from its little pocket. As it does so it becomes hardened. The visible section of the hair, which gets longer and longer, is called the hair shaft. The shaft increases in length by about a third of a millimetre a day.

The speed at which our hair grows varies with age and health. It grows most slowly in old age, illness, pregnancy and cold weather. It grows quickest during convalescence after a serious illness, apparently as a compensation mechanism after a period of impeded growth. In healthy people it grows fastest when we are between the ages of 16 and 24. During this period, the annual increase is up to 7 inches a year, compared with an overall average of about 5 inches.

The life span of an individual hair is about six years, which means that, for a healthy young adult, an untrimmed hair would grow to roughly 42 inches in length, before falling out and being replaced. This means that straight-haired young men and women who never cut their hair would have flowing locks reaching to their knees. Running around naked in such a condition, back in our primeval days, we must have been one of the strangest-looking creatures in the animal kingdom.

Apart from its inordinate length, another oddity of our hair is that it does not show a seasonal moult. We could easily step up our hair loss at the start of the summer and slow it down in winter, to create a more finely tuned insulation layer, like many other animals, but we show no signs of this. Each of our individual hairs has an independent lifecycle. At any one moment, 90 per cent of the scalp hairs are actively growing and 10 per cent are 'resting'. The resting ones are scattered throughout the others and they remain in this inactive state for about three months before dropping out. This means that we each lose between 50 and 100 head hairs every day.

When a hair drops out, both the long shaft and the short root become detached, but the tiny papilla at the base of the follicle stays put. This little bud then starts to sprout a new hair to replace the old one. It remains active for nearly six years, when it once again stops growing

▲ Although baldness is common, the opposite condition is extremely rare. Hairy-faced people do, however, occur from time to time, giving us a glimpse of what our ancient ancestors might have looked like before they lost their coat of fur and became 'naked apes'. Until quite recently such hairy people could earn a living as circus freaks, but nowadays they have the benefit of modern skin-depilation techniques.

▼ The strength of healthy human hair is remarkable. Chinese circus acrobats have been known to perform tricks while suspended by their hair, without undue discomfort. A single strand of Mongoloid hair is reputed to have a breaking strain of 160 grams. It is also highly elastic, and can be stretched by as much as 20 or 30 per cent before snapping.

new cells and becomes dormant. After another three-month resting period it sheds its old hair and repeats the whole process. With our extended lifespan today, each papilla can be expected to repeat its lifecycle about 12 times, creating 12 complete hairs, one after the other, each hair being several feet in length. It follows from this that if an individual human being lacked the dormant phase in the hair cycle, he or she would be able to grow locks of hair up to 30 feet in length. This freak condition does seem to have occurred at least once. Swami Pandarasannadhi, an Indian monk from a monastery near Madras, was reported to have unkempt hair which stretched for 26 feet. There are also a number of other, less spectacular claims of women whose hair was longer than their bodies.

The opposite effect, a permanent switching-*off* of hair-growth, is much more common. It never occurs during childhood, but when sexual maturity arrives strange things start to happen on top of the male head. The male hormones flooding through the system de-activate certain selected hair papillae. The ones around the sides of the head are spared, but those on the crown of the head are knocked out of action. When the individual hairs fall out they are not replaced. The papillae, instead of becoming dormant for three months and then starting up again, become dormant forever. As a result their owner becomes bald.

Balding is usually a gradual process, and many men escape it altogether. About one in five starts to go bald soon after adolescence, although the change is so slight at first that it is hardly noticed. By the age of 30, however, the balding 20 per cent will have become well aware of what is happening. By the age of 50, some 60 per cent of white males will have shown some degree of hair loss. (The figure is lower for other races.) Elderly females also show a slight hair loss, but of a different kind. In their case, there is an overall thinning of the hair which is far less noticeable, and which contrasts strikingly with the 'bald patch' pattern of the males.

One way to prevent baldness is to have yourself castrated before you reach puberty. There were no bald eunuchs in the Sultan's harem. Another way is to have yourself born into a family where all male ancestors were fully haired into their dotage. If the baldness genes are completely missing from your family tree you will never need a toupee. You might have to pay a small price for this condition, however, because your balding friends will probably have a stronger sex drive than you will. It has been shown than an over-production of male sex hormones (androgens) is a vital factor in producing a bald head. Bald may not be handsome but it is almost certainly sexier.

Because it is linked with high levels of sex hormones and because it increases in extent with advancing age, it is obvious that baldness is a human display signal indicating male dominance. It typifies the virile older man and separates him visually from two other categories, the virile younger male and the less highly sexed older male. There appears to be a flaw in this arrangement, however. As men become elderly they all experience a waning of their sex drives, bald or hairy. Bald men may lose their sexual urges more slowly than the rest, but lose them they do. Logically, they should then start to sprout hair again on top of their heads, but this does not happen. Their gleaming-pate display becomes a cheat. It seems as if, after years of

■ It has been argued that the differences in hair pattern between the major human races are due to adaptations to climate. The crinkly hair of the Negroid race is thought to he a specialized protection against the hot sun, for example. But there are difficulties with this interpretation when climatic explanations are sought for the differences between the wavy hair of Caucasians, the straight hair of Orientals and the peppercorn hair pattern of Bushmen (and some of their mixed-race relatives).

■ Human hair colour ranges from pure black to pure white with a varying degree of additional red. This simple combination gives rise to all the well known browns, auburns and blonds in addition to the more dramatic blacks, reds and albinos. The significance of red-coloured hair remains something of a mystery. It reaches its highest level of incidence in the border regions of southern Scotland, but it also crops up in many other places without any apparent advantage to its owners.

inactivity, the hair papillae are no longer capable of being revived. But all is not lost. A new signal is added at this point which converts the dominant virile male image into that of a 'grand old man': his hair turns white. This happens to both hairy and to bald heads and transmits the vital 'I-am-old' signal in both cases, so that the bald pate of the elderly male does not cheat after all.

Before considering social attitudes to hair, there are a few remaining anatomical details to record. Yet another oddity of the human hair pattern is that it lacks 'feelers', or vibrissae, the tactile hairs we know so well in their familiar form as cat's whiskers. All mammals, even whales, have at least a few vibrissae but, for some reason, we alone are without them. Nobody knows why. Another missing element is the ability to make our hair stand on end when we are angry. Many mammals are capable of bristling with rage and making themselves appear much larger in the process, but we have lost this dramatic transformation display. This is not so surprising. Erecting our short sparse body hairs wouldn't frighten a mouse, and our head hairs are obviously too long and too heavy to be hoisted into the angry-erect condition.

Despite the lack of hair-erection displays we do, however, still retain the tiny muscles which move the hairs. Called the *arrector pili* muscles, the best they can do for us today is to give us goose pimples when we are frozen with cold or fear. What they are doing is making a pathetic attempt to 'thicken' our non-existent coat of fur. If we still

▶ We speak of 'grey hair' but there is no such thing. Hair does not turn grey; it turns white. 'Grey' hair is simply a mixture of old hairs, which are still their original colour, and new hairs scattered among them, which are pure white. At first the white hairs make no visual impact, but as they increase in proportion to other hairs they give the overall impression of greyness. Eventually, as more and more follicles give up pigment production and create unpigmented hairs, these white strands swamp out the few remaining coloured ones and at last replace them altogether, leaving a shock of white hair to signal that old age has arrived.

had our fur, this thickening would increase the insulating layer of entrapped air and in this way would help to keep us warm.

Having stated this general rule about the inefficiency of our hair-erection, it is necessary to mention an exception. It concerns the re-action to the sound of a creaking door, late at night in a dark house. 'It made my flesh creep' is the usual description, and this creeping sens-ation is, of course, the thousands of *arrector pili* muscles contracting. Sometimes people say 'My hair stood on end' and they comment that the sensation was strongest on the back of the neck. This is probably because it is there that the hairs are dense and short enough to be able to produce a particularly strong local response.

One way in which our human hairs do not differ from those of our mammalian relatives is in the presence of associated sebaceous glands. These tiny glands, situated at the side of the hair root inside the follicle, produce an oily secretion, the sebum, which helps to lubricate the hairs and keep them in good condition. Over-active sebaceous glands produce greasy hair, under-active ones give rise to dry hair. Hair-washing is important to remove dirt from the hair, but it also removes the natural sebum, and too much washing can be almost as damaging as too little.

The strength of healthy human hair is remarkable, as anyone who has been dragged along by it will have discovered. Chinese circus acrobats have been known to perform tricks while suspended by their hair, without undue discomfort. A single strand of Mongoloid hair is reputed to have a breaking strain of 160 grams. It is also highly elastic, and can be stretched by as much as 20 or 30 per cent before snapping.

Hair colour varies in a rather simple way, largely following the colour of the skin. The same pigmentation system is used in both cases. People living in sunny countries have a large number of elong-ated melanin granules in the cells of their hairs, making them appear black. People in more temperate zones have slightly less melanin, giving their hair a brownish colour. Up in the cooler, sunless world of Scandinavia, there is even less melanin, resulting in the pale-colour-ed hair we call blond. Albinos, who lack melanin completely have pure white hair.

This simple scale of black to white is complicated by one 'rogue' element. Certain individuals have melanin granules which, instead of being elongated, are spherical or oval in shape. These are seen by the eye as red. They may appear by themselves, with none of the usual elongated melanin granules, in which case their owner will appear as a 'golden-blond'. If they appear in combination with a moderate amount of the ordinary granules, then their owner will have a rich, reddish-brown tinge and will be referred to as a 'fiery redhead'. If the spherical granules occur in combination with large numbers of elongated granules, then the blackness of the hair will almost mask the redness, but it will still be present to give a subtle tinge to the hair and make it different from the pure black variety.

The shape of each hair varies considerably, but three main hair-types are now recognized: the crinkly (heliotrichous), typical of the Negroid race; the wavy (cynotrichous), typical of the Caucasian race; and the straight (leiotrichous), typical of the Mongoloid race. It is usually claimed that these three racial hair-types are related to clim-

Long hair may give an assailant a handhold.

atic conditions, but several objections have been raised against this view. It is accepted that the tightly looped hairs on the Negroid scalp do create a bushlike barrier between the skin and the outside world, where the sun is beating down viciously on top of the human head. This creates a buffer zone of entrapped air which helps to prevent overheating of the skull. But if such a buffer zone is so efficient in hot climates, it is argued that it would be equally valuable in cold climates, where the entrapped air would act as an insulating device.

Looking at the wavy hair of the Caucasians, two more problems arise. To start with, Caucasian hair is very variable, ranging from almost straight to curly, without any reference to shifts in the environment. Secondly, Caucasians live successfully in a huge range of habitats, from the frozen northlands of Scandinavia to the searing heat of Arabia and India. A similar problem arises with the straight hair of the Mongoloids. It may show little variation, but it has an even wider range from north to south – from Siberia to the East Indes and from Alaska to the Amazonian jungles of South America.

The main argument to set against these objections is that recent migrations of human beings have spoiled the original pattern. Suppose, for a moment, that in the beginning there were crinkly-haired people in hot countries, wavy-haired people in temperate countries and straight-haired people in cold countries. The crinkly hair would combat the overhead sun without hanging down to prevent sweating from the neck-and-shoulder region. The long, straight hair would act like a cape over the neck and shoulders, keeping them warm. And the wavy hair would be a compromise between these two extremes, suitable for the intermediate zone. Then, from this starting point, huge migrations take place, too swift for genetic alterations in hair-type to keep up with them.

This scenario makes sense, but it is little more than guesswork. The fact that the three main types of hair *look* different may also have been playing a role. If at some point in the distant past the three major races of mankind were pulling apart from one another, they may have used visual differences of this kind as isolating mechanisms and the three hair-types may have been important as racial 'flags'. This may have kept the three types going even after migrations had taken their owners into inappropriate climatic conditions.

Up to this point hair has been considered in its natural condition, but the busy fingers of humanity have rarely left it in that state. The urge for self-decoration has produced some startling modifications and distortions. The most basic and widespread attack has been to interfere with the natural length of the hair.

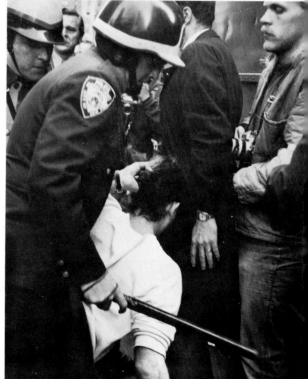

Hair length alterations have nearly always been linked to the introduction of a new gender signal. As already explained, male and female hair in the natural state shows little difference in this respect, so that it is quite arbitrary as to which sex gets the 'short straw' in any given culture. For some tribal societies, the males enjoy elaborate hairstyles while their females display shaven heads. In other cultures, the long tresses of the females are their 'crowning glory', while the males clump about in bristly scalp-stubble.

As a result of these hair-length contradictions, two completely different types of hair symbolism are interwoven in human folklore. In one strain, the male's great mop of hair is seen as his strength and his,

■ Although male head hair is biologically as long as female head hair there is an old tradition founded on St Paul's words to the Corinthians which sees closely-cropped hair as essentially masculine. This is probably supported by the aggressive male's fear of being grabbed by his long locks in a fight. Police often use this handhold when arresting rioters. A crewcut or skinhead male robs them of this advantage. The long tresses of the female, on the other hand, give her an air of vulnerability – she is symbolically available for hair-pulling, pinioning or dragging off to a cave.

▼ The forcible shaving of a female head in a culture where long female hair is the norm has been used as a public punishment as recently as 1944. In Paris at the close of World War II women accused of collaboration with the German occupying forces were taken into the streets and there clipped and shaved in the centre of a hostile crowd. This type of mutilation, although bloodless, is effective because of the long time it takes for the hair to re-grow and remove the badge of shame. In some African tribes, such as the Masai and the Dinka, this form of punishment would be meaningless because in those cultures all females have shaven heads as an everyday hairstyle. For them the appropriate punishment would be to forbid head-shaving.

virility, giving him power, masculinity and even holiness. The word 'Caesar', for instance, (and its derivatives Kaiser and Tsar) meant hairy or long-haired and was thought eminently suitable for great leaders. This tradition goes right back to the earliest of hero figures, the Babylonian Gilgamesh, who was both long-haired and immensely strong. When he grew sick and his hair fell out, he had to go on a lengthy journey so that 'the hair of his head was restored' and he could return refreshed, with his mighty strength renewed.

In this folklore tradition, the virility of the male's copious head of hair is undoubtedly linked to the fact that, although sex hormones make both male and female bodies hairy at puberty, it is the masculine body that becomes hairier, sprouting not merely pubic and armpit hair, but also beard, moustache and, frequently, straggling hairs on trunk and limbs. If extra-hairiness is masculine, then *all* hair becomes symbolic of masculine power and virility, even the hair on the head.

It followed from this that to shave the head of a male was to humiliate him and that to shave one's own head was a sign of humility. For this reason, many priests and holy men cropped their heads to humble themselves before their deity. Oriental monks went shaven headed as a symbol of celibacy. Inevitably, psychoanalysts have interpreted the cropping of male hair as displaced castration.

This whole tradition was completely contradicted by no less a person than St Paul, who told the Corinthians that it was natural for a man to have short hair and a woman to have long hair. He seems to

Some hairstyles are are intended to impress or outrage; a few are actually dangerous.

▼ *In ancient times long male hair was considered a source of masculine strength. When Delilah cut Samson's hair she robbed him of this source of power. Today manly attributes are more often associated with short hair, and young males who sport long flowing locks – especially if styled in the fashion of the opposite sex – are consciously exploiting the 'outrage factor' guaranteed by a prevailing tradition of hair symbolism.*

▲ *Before there were written records it was the tribal elders who were the source of wisdom and the keepers of the tribal secrets. Today the role of the elderly male as a fount of knowledge is greatly reduced in importance, but the bald yet hirsute patriarchal figure with his great dome, white hair and flowing white beard, still makes an impact.*

have been influenced in this decision by the Roman military custom of cropping soldiers' hair. This appears to have been done, not as a humiliation, but simply as a means of increasing uniformity and discipline and to make the Roman troops look different from their long-haired enemies. An element of hygiene may also have been involved. Whatever the reason, St Paul came to the slightly dotty conclusion that short male hair was a glory to God, while long female hair was a glory to man. For this reason he demanded that men should always pray bare-headed, while women should always cover their heads in prayer – a Christian custom that has lasted for two millennia, despite the fact that it is based on a complete misunderstanding about human hair.

St Paul did not mince his words. At one point he said: 'Does not even nature itself teach you that if a man has long hair, it is a shame unto him? But if a woman has long hair it is a glory to her; for her

■ Once outrageous hairstyles have become
commonplace it becomes a problem for a social
rebel to find a new way of appearing unorthodox.
Even extreme fashions have to be surpassed. One
unique punk solution was to apply superglue to
the hair, training it up into hard sharp spikes.
The owner of one such hairstyle was sacked from
his factory job because his porcupine fashion
'represented a safety hazard. The spikes projected
from his head for some distance and could have
injured a supervisor.'

▶ The resistance of
the human animal to
appearing dowdy and
dull has always been
strongly expressed in
tribal societies where
young males in particu-
lar may spend many
hours decorating them-
selves and especially
their hair. Even in the
concrete deserts of
modern cities rebellions
still occur against bland
uniformity, and in the
1980s there has been
a striking resurgence
of dramatic hairstyles of
astonishing unlikeliness
and impracticality.

Males have devised many ways of concealing baldness.

hair is given to her as a covering.' Although there have been many attempts to break away from this dictum of St Paul's, its influence is still with us to this day. Despite occasional hirsute rebellions by Cavaliers and Hippies, the shaggy, long-haired male has remained a rarity, and despite similar rebellions by bobbed and snipped modern females, the short-haired female has also proved to be a rare species during the centuries since Paul laid down his bizarre ruling. It is as though, deep down, we have come to believe him and to accept the false premise that short hair is, somehow, more masculine.

One possible reason for this may relate to the harshness or softness of the hair. Closely cropped hair is bristly and rough. Long flowing tresses are soft and silky. Rough and soft – male and female – could be an unconscious factor tending to promote an acceptance of artificially-shortened male hair. Even in these days of sexual equality we still seem to be unable to return to the natural state of hair-length equality.

A special problem faced by the long-haired female has been that her soft flowing hair with its silkily erotic texture has often been too provocative for sexually inhibited societies. Puritans hated its sensuality, but were unable to demand its cropping because this would have made it unfeminine and contrary to the law of God as laid down by St. Paul. The solution was simple – the hair had to remain long but had to be hidden from view. This meant stuffing it under a close-fitting cap or some other concealing form of pious headgear, or, at the very least, tying it up in a tight bob.

Because this made long female hair a rare sight in public, the act of 'letting her hair down' became a powerful moment of intimacy between a woman and her husband, and the unpinning of a cascade of tumbling locks took on a strongly erotic flavour. In certain epochs, to wear the hair loose in public became the hallmark of a 'loose woman'. One of the public punishments for a woman of easy virtue was to shave her head and remove her 'badge'. This was done as recently as 1944, when Parisian women accused of sleeping with the German occupation forces were head-shaved in the streets, daubed with swastikas and paraded through the city.

Returning for a moment to the question of the balding male, it is important to realize that his condition has little bearing on the long-hair/short-hair debate. A bald man may still present a long-haired or short-haired appearance, depending on whether the surviving ring of hair around his bald patch is allowed to grow and hang down around his head or is closely cropped. So it makes little difference to him which trend he follows. Something else does seem to matter to him, however. In modern times he has often gone to great lengths to conceal his baldness from the eyes of his companions. Bearing in mind earlier comments explaining that baldness reflects an unusually high level of male sex hormones, this curious reluctance to display his virility requires some explanation.

The answer has to do with the strange delay in the development of the bald pate. Although many men start to lose their hair while still in their twenties, the effect is usually minimal. It is not until they reach middle age that the expanse of bald skin becomes glaringly obvious. In old age, it continues even further, of course, so that baldness becomes, not so much a symbol of male virility as of male aging. In a culture which worships youthfulness this is clearly something of a

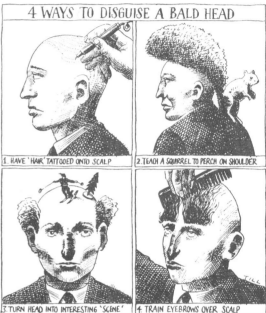

4 WAYS TO DISGUISE A BALD HEAD

1. HAVE 'HAIR' TATTOOED ONTO SCALP

2. TEACH A SQUIRREL TO PERCH ON SHOULDER

3. TURN HEAD INTO INTERESTING 'SCENE'

4. TRAIN EYEBROWS OVER SCALP

■ *Many cures for baldness have been offered in the past, some cunning, some brutally surgical and some merely fanciful. Countless lotions, oils and greases have been sold to hopeful customers over the years, but all have failed. Only within the last few years has a substance arrived on the scene which at last offers a slender hope. Called* minoxidil, *its hair-growth potential was discovered by accident. It was being used to control blood pressure and in this role was administered to a man who had been bald for 18 years. Within four weeks of the treatment normal dark hair was growing on top of his head. The excitement over this amazing side-effect was slightly dampened by the fact that he was also sprouting hair on his forehead, his nose, his ears and other parts of his body. Switching to a locally applied solution containing minoxidil, doctors then started rubbing the bald heads of selected patients and found that with certain kinds of baldness they had an 80-per-cent success rate, even though the hair growth was rather modest. Other chemicals are now under serious test and the baldies of future generations may well be spared the traumas of bloody hair transplants to replace the fur on their shiny pates.*

■ Apart from surgical and chemical ways of concealing baldness there are several less demanding solutions. One is to sweep the remaining hair up and across the bald spot, using a lower and lower parting as the years go by. Another is to adopt some kind of speciality headgear. And finally there is that ancient standby – the wig, hairpiece or toupee. In ancient Egypt the pharoahs and their families shaved their heads and then covered them with ceremonial wigs. Slaves were forced to wear their own hair, by law. This high-status role for wigs was repeated elsewhere. Wigs are known to have been worn by Assyrians, Persians, Phoenicians, Greeks and Romans. It was in 17th and 18th-century Europe, however, that stylized wigs reached their zenith. Although they were introduced during this period as a means of concealing baldness, they soon spread to become a fashionable mode of dress for all ranking members of society and were even worn by schoolchildren at top boarding schools. By the 1750s the fashion was dying and today survives only in the antique atmosphere of the courtroom. Except in the theatre, wigs nowadays apper only as secretive, realistic hairpieces.

disaster, especially for males who appear in public as actors or singers who are meant to be transmitting sexual charm during their performances. If 18-year-old males are at their highest sexual peak (which they are) and if 18-year-old males never display bald heads (which they don't), then older males must aspire to copy the 18-year-old condition in as many ways as they can. For professional performers reaching middle age this may require rigid dieting and, above all, the display of a hairy head.

If the hairs desert such males, frantic efforts may be made to correct the situation. There are five main courses of action. The first is to rub into the shiny scalp some elixir of youth. The second is to undergo a savage surgical procedure in which hairy tissue is transplanted on to the bald patch on top of the head. Far less drastic is the 'Sidewinder' technique. All this involves is a careful combing and brushing of hair from the side of the head, so that it spreads over and covers the bald pate. A fourth method, much favoured by bald actors who enjoy outdoor sports, is the wearing of exotic hats to cover their gleaming shame. Finally there is the popular old stand-by, the wig, hairpiece or toupee.

In addition to these five there is one spectacular, but rarely-used ploy which operates in a totally different manner. This is the device employed by Yul Brynner, Telly Savalas and a small band of other brave men who shave their entire heads, and keep on removing every scalp hair, day after day, so that it is impossible to distinguish a bald 'patch' on top of their heads. By scraping away the scalp hair that would normally survive baldness, they give the impression that they have deliberately chosen to do away with *all* head hair. 'I am not bald,' they are saying; 'I am shaved, and that is my choice.' In terms of association this puts them into one of several categories – humble oriental monks, ancient rulers, sheared criminals, or professional wrestlers. Their dominant, active life styles narrow this list down, so that they emerge with the personalities of 'wrestler-kings' – tough men who scorn orthodox fashions and who appear dignified, yet ready for a fight. Compared with the sneakiness of the toupee-wearers, there is something brave and swashbuckling about their blatant defiance of the laws of hairiness, and they come out easy winners. Whether they would be so successful if they were less rare is, however, another matter.

Turning to the general question of hair decoration, it is safe to say that there is no society or culture anywhere on the globe today which does not decorate or style the hair in some manner. And this has been the case for literally thousands of years. Hair has been dyed, shaped, lacquered, curled, straightened, powdered, bleached, tinted, waved, braided, dressed, greased and oiled in a million different ways at a cost of countless man-hours of human labour and ingenuity. One of the reasons for the special attention paid to this part of the human anatomy is that hair can be altered in so many ways and, above all, because, after it has been cut or modified, it eventually grows back again. This constant renewal of scalp hair has made it into a suitable symbol of the life force itself and has loaded it with a huge variety of superstitious beliefs and taboos.

The giving of a piece of hair in a locket to a loved one was an act of total surrender to him or her. Far more than a sentimental, loving act

The completely bald or shaved head is sometimes seen as an assertion of male dignity and strength.

■ The cropping of the head has been associated with masculinity ever since St Paul decreed that short hair is natural for the male. It follows that extreme cropping – in other words, shaving the head – is super-male and super-virile. This device has been used by sportsmen and by certain actors with dominant personalities. For wrestlers and warriors it has the added advantage of robbing the adversary of a convenient hair-hold. And for swimmers it improves streamlining in the water. In tender moments, at close quarters, it also provides an appealingly kissable surface reminiscent of a baby's bald head.

► The enforced shaving of the head has always been seen as an act of humiliation. In ancient times male hair meant strength and its removal (as in the case of Samson) meant a destruction of that strength. The shaving of the head of a prisoner, slave or criminal was an act of degradation – a mild type of body mutilation that has been seen as a symbolic form of castration – and voluntary headshaving was restricted to individuals who wished to make a public display of humility towards their gods. Oriental monks still follow this ancient practice, but their message of humble celibacy is easily confused today with the virility symbolism of the other modern shaved heads seen here.

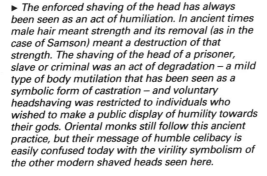

▲ *The action of scratching the back of the head has an intriguing origin. It is done in moments of frustrated aggression and is derived from our primeval attack movement. When angered and about to strike someone, we automatically raise the arm to deliver a downward blow. The frontal blow of the trained boxer is a much more sophisticated movement and has to be learnt, but even tiny children doing battle in the nursery employ the overarm blow, and it stays with them throughout their lives. If as adults they become involved in street-rioting they will revert to it again, and the riot police will respond in a similar manner, beating them over the head with clubs. A man in an angry but inhibited mood at a social gathering cannot thump the person who is annoying him, but his arm flies up as if to do so, responding to the primeval prompting of his unconscious mind. When the arm reaches the uppermost position, ready to start on its downward arc, it is checked and the impotent hand diverts itself with a vigorous scratch or pat, as if to suggest that that was the intention all along.*

▼ *Attention to hair-grooming reflects the personality of the groomer. Tight immaculate grooming without a hair out of place indicates a tightly controlled, powerfully self-disciplined individual. A loose, free-flowing style suggests a more vulnerable, open spirit. It was because of this that Victorian women wore their hair pinned up during the day and only 'let their hair down' in the privacy of their bedrooms.*

to help recall your presence, it was a symbolic placing of your soul in the other's power. The lock of hair contained the vital spirit of the giver and, by wearing it around the neck, the loved one was given the power to control and bewitch the donor. An unusual variant of this concerns the chivalrous knights of the Middle Ages. Dedicated to courtly love, these brave warriors wore a tuft of their mistresses' pubic hair in their hats when they went into battle.

Because of its magical powers, barbers in superstitious societies were forced to bury the shorn hair of their customers in secret places so that it could not be stolen and used in magical ceremonies to harm them. Such customs are by no means dead today. Parents in rural districts of Europe are still warned not to keep locks of their children's hair if they wish them to live a long life. This is because, once again, it is feared that 'evil beings' may get hold of the clippings and use them to cast a spell on their owners.

Among gestures involving the scalp hair is that of the female who repeatedly rearranges her hair with her hands as she talks to a male friend. She may flick the hair up, as if to loosen it from her collar, or she may comb it by running her fingers through it, or start brushing it backwards with her hand. All these small movements are unconscious grooming movements which indicate that the female is unwittingly saying 'I want to improve my looks for you'. Without realizing it she is transmitting powerful invitation signals to encourage the male.

In ancient times, a more violent hair-contact signal – the tearing of one's hair – was a common gesture of mourning and despair. In extreme cases, women pulled out whole tufts of their hair and scattered the strands over the corpse, as a demonstration of sorrow.

People rarely touch one another's scalp hair unless they are lovers, parents or hairdressers. The head region is a 'protected area' and is not available to casual acquaintances, largely because of its close proximity to those precious and extremely delicate organs, the eyes. Only the most highly trusted companions may make contact with the head hair. When this happens a number of characteristic actions are observed. There is the 'laying on of hands' when a priest blesses a believer. There is the pat on the head of a child by a proud parent, or the mocking pat on the head between adult friends suggesting condescendingly that one of them is behaving like a little child. There is the intimate hair-to-hair contact of lovers who put their heads together and, later on, the hair-stroking, fondling and kissing of love-making.

Above all, there are the many hours of life spent having the hair groomed by the professional hair-touchers – the barbers and hairstylists. This goes far beyond the demands of cleanliness and even beyond the need for decoration and display. It harks back to those distant primitive times when, like our close relatives the monkeys and apes, we spent lengthy periods of each day grooming one another's fur. As with all primates, this activity was much more than a comfort activity – it was a method of cementing social friendships within the group. It was caring, non-aggressive physical contact with another being and this gave it a deeply rewarding feeling. It is the same pleasant sensation that people feel today, millions of years later, when they give themselves up to the luxury of the hair-groomers' hands.

THE BROW

■ To have a brow like a human being you have to be a very intelligent animal indeed. For the human brow, made up of forehead, temples and eyebrows, was the direct result of our ancestors' dramatic brain enlargement. The brain of a chimpanzee has a volume of roughly 400 cubic centimetres; for modern man the figure is 1,350 – more than three times that of our hairy relatives. It was the expansion of the human brain, especially in the frontal region, which gave us a 'face above our eyes'.

If you take a close look at a chimpanzee's face, side by side with a human face, the forehead-difference is striking. In the case of the ape it is almost non-existent. In the human it rises vertically above the eyes as a great naked patch of skin. The chimpanzee's hairline comes right down to the eyebrows, which are almost hairless. In fact, the brow region of the ape is the complete opposite of that in human beings.

The difference in brow-ridges also demands explanation. In apes they are bony crests over the eyes which protect them from physical damage. The early ancestors of man also possessed these heavy rims of bone, but they gradually dwindled in size until, with us today, they are almost gone. Why did we lose them? When we became primeval hunters we would surely need them even more than in our remote fruit-picking days.

The answer is that their loss is more apparent than real. If profiles of the head of an ape and of a man are compared it emerges that the protective line of the brow-ridge remains roughly where it is, while the forehead expands above it. By the time the condition of modern man has been reached, the forehead, inflated by the ballooning human brain, has pushed forward to the level of protrusion of the ancient brow-ridge. So the new forehead acts in the same protective way, providing a bony defence against blows to the eyes. The brow-ridges have not vanished, they have simply been engulfed.

It could be argued that, with the increased hazards of violent hunting activities, it might have been an advantage to have doubled the protection, to have kept the bony ridges in addition to the bulging forehead. One suggestion as to why this did not happen is that during the Ice Age our shivering forbears developed more-flattened faces as a defence against the cold. The move towards flattened fat-lined faces of the kind we still see today on Eskimos included a reduction of the sinuses over the eyes, sinuses which became vulnerable to infection in the colder climate. This reduction had the effect of flattening the brow.

The difference between the eyebrows of ape and man is also intriguing. In both cases the main evolutionary effort appears to have been to make them conspicuous and contrasting with their surroundings. A young chimpanzee has pale, naked eyebrows which stand out vividly against the dark hair above them. A young human has dark eyebrows which stand out against the pale skin above. Even in dark-skinned races the contrast is not lost, so that movements of the eyebrows are still clearly visible to those nearby.

There is little doubt that the function of these conspicuous 'superciliary patches', as the eyebrows are known technically, is to signal the changing moods of their owners. Once, it was thought that their main duty was to prevent sweat and rain from dripping into the eyes. Although they may be of minor assistance in this way, acting as de-

■ The chimpanzee has a shallow sloping hairy forehead with naked eyebrows and large brow-ridges. The human has a deep vertical naked forehead with hairy eyebrows and small brow-ridges.

▼ The loss of protective brow-ridges during human evolution is more apparent than real. This becomes clear when gorilla and human head profiles are compared.

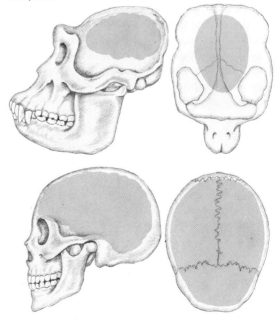

The main function of the eyebrows is to signal the changing moods of their owner.

◄ *The bushier eyebrows of the human male help to make his brow movements more expressive, emphasizing his frowns and grimaces at moments of emotional tension.*

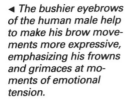

► *The visual signals transmitted by the postures of the eyebrows are amplified by conspicuous brow-wrinkles. These furrows become increasingly marked as advancing age reduces the elasticity of the brow skin.*

► *Elderly people with weather-beaten faces display permanent wrinkling of the forehead. It is almost as though each individual had a giant 'finger-print' impressed on the brow.*

■ *The two young girls seen here eloquently demonstrate the signalling power of the human eyebrows, showing the marked contrast between the arched eyebrows of greeting or pleasant surprise and the oblique eyebrows of distress.*

flecting 'gutters' at the bottom of the forehead, their major value is undoubtedly connected with facial expressions. Every time our mood alters, the position of our eyebrows changes, producing a variety of important brow signals, as follows:

1. The Eyebrows Lower. This action, the frown, is not strictly vertical. As the eyebrows move downwards they also move slightly inwards, coming closer together. This has the effect of squeezing the skin between them and throwing it into short, vertical folds. The number of these folds varies from individual to individual and each adult has a characteristic 'frown pattern' of one, two, three or four lines. Frequently they are asymmetrical, with the lines on one side of the inter-eyebrow space (an area known technically as the glabellum) being longer or stronger than the other.

The horizontal lines on the forehead tend to be smoothed out by the action of lowering the eyebrows, but they may not disappear completely. Part of the aging process in the human animal involves an increasing fixation of the temporary expression lines. Skin-

▲ *It is often imagined that the region at the sides of the forehead are called the temples because this part of the human body houses the mystical powers of great thinkers and artists, but this is not the true reason. The temple is named from* tempus, *meaning* time, *because it is here that the beating of the pulse is visible on the skin surface.*

creases which, in the young, appear and vanish again with each changing mood, become permanently etched into the skin surface as the years go by. The strength of a frown-line on a non-frowning face is a fair indication of the sum total of past frowning performed by that individual.

Eyebrow lowering occurs in two quite different kinds of situations which can be crudely labelled as aggressive and protective. In aggressive contexts the action covers a wide range of intensities, from mere disapproval or self-assertive determination right through to annoyance and violent anger. In protective contexts the action occurs whenever there is a threat to the eyes.

At moments of danger the lowering of the brows over the eyes is not protective enough, however, and in such instances there is also a raising of the cheeks from below. Together these two actions provide the maximum of eye protection that is possible, while keeping the eyes open and active. This 'screwing-up' of the eyes is typical of a wincing face that anticipates physical attack, or of an over-illuminated face exposed to such strong light that the eyes are beginning to suffer from it.

This protective 'screwing-up' of the eyes also occurs frequently during laughter, crying and at moments of intense disgust, indicating that these conditions can also perhaps be looked upon as kinds of 'over-exposure'.

It is the eye-protection function that explains the ancient origin of the lowering of the brow region. The use of this lowering in aggressive contexts appears to be secondary, based on a need to defend the

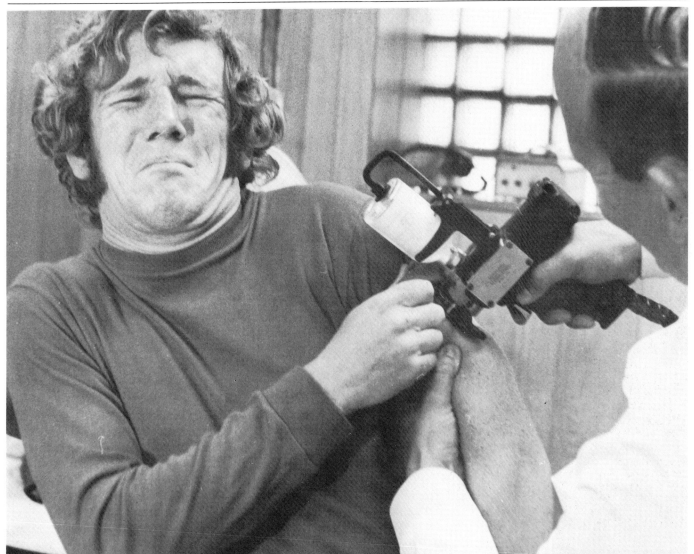

▲ The lowering of the brows is part of the ancient eye-protection reaction to real or imagined danger. If we sense that something is about to cause us pain we respond with a 'wince' movement in which the eyes are instantly surrounded and buffered by lowered brows and raised cheeks.

eyes from the retaliatory attacks which the aggressive mood can be expected to provoke. We often think of the frowning face as a 'fierce' face and therefore not one that should be connected with self-protection, but this would appear to be an error. It may be fierce but it is not so fearlessly fierce as to be unconcerned about the need for self-protection of those crucially important organs, the eyes. The truly fearless face of aggression, by contrast, displays a pair of staring, unfrowning eyes, but this is a comparatively rare occurrence since overtly hostile acts are seldom safe from some kind of retaliatory response.

2. The Eyebrows Raise. Like the last movement, this action is not strictly perpendicular. As the eyebrows are raised they move slightly outwards, away from each other. This has the effect of stretching the skin between them and flattening out the short vertical frown-folds there. At the same time, however, the whole of the forehead skin is squeezed upwards, creating a pattern of long horizontal crease-lines. These are roughly parallel with one another and number four or five in most cases. Sometimes there are as few as three or as many as ten, but it is difficult to be precise because the upper and lower lines

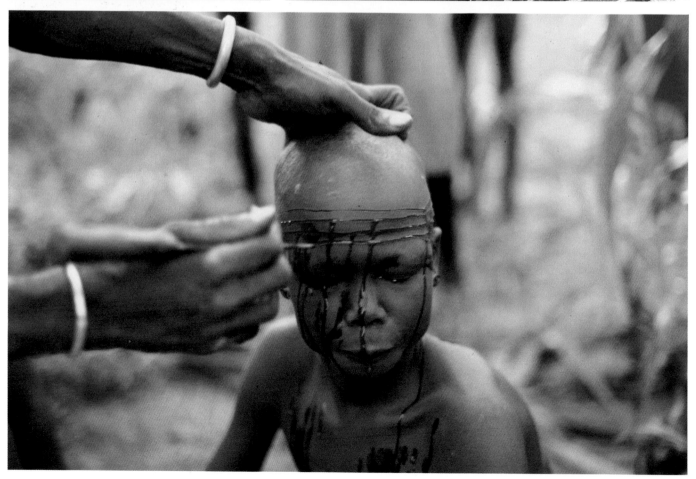

■ In modern industrial societies the decoration of the brow region is limited largely to female eyebrow plucking and pencilling, but in tribal societies more drastic modification can occasionally be found. Certain tribal initiation ceremonies involve cutting patterns of lines across the forehead to create a tribal 'label' for the initiate. In later life such incisions produce a fixed brow expression regardless of the true feelings of the owners. An even more dramatic technique is employed in parts of the Sudan, where a row of small cuts is made above the eyebrow line into which small stones are inserted, creating a striking pattern of giant 'warts'. For most tribal societies, however, the decoration of the forehead is limited to costume adornments, with colourful head-bands and other easily removed ornaments. The brow region of the human body is too expressive to be a popular zone for permanent adornment involving skin disfigurement.

are usually fragmentary. Only the middle lines run right across the brow in most instances.

This is the 'furrowed brow' of popular speech and is usually thought of as belonging to a 'worried' individual. Its true range, however, is much wider than that. It has been described by various authors as signifying: wonder, amazement, surprise, happiness, scepticism, negation, ignorance, arrogance, anticipation, querying, incomprehension, anxiety and fear. The only way to understand its significance is to look backwards towards its ancient origin.

Brow-raising is a pattern we share with other primate species and for them like us, it appears to have started out as a vision-improving device. The pulling up of the forehead skin and the raising of the eyebrows has the immediate effect of increasing the overall range of vision. To use a well-known phrase, it is an 'eye-opener'. It increases visual input.

Among monkeys it seems to be an emergency reaction, brought into use whenever the animals are confronted by something that makes them want to flee. It only occurs, however, if, at the same time, there is something else in the situation which stops them from fleeing. This 'something else' can be a number of things. It can be a conflicting urge to attack, or a burning curiosity to stay and look at the thing that is so frightening, or any other tendency to stay put which conflicts with the urge to flee and blocks it.

If we now apply this concept of 'thwarted escape' to the human context it fits remarkably well. Man and monkey behave in much the same way. The worried man with the furrowed brow is essentially a man who would like to escape from the situation he is in but for some reason cannot do so. The laughing man with the same brow-wrinkling expression is also slightly alarmed. There are tell-tale elements of body withdrawal in his posture. His laughter may be genuine but whatever it is he is laughing at is also rather disturbing. This is not uncommon. Much humour takes us to the brink of fear and makes us laugh only because it does not push us over. The arrogant man with his stiffly arched eyebrows would also like to escape – from the crassness which surrounds him. In popular speech, we give him the appropriate name of 'high-brow'.

When this expression is compared with the lowering of the eyebrows, a problem emerges. Suppose we see something frightening in front of us: we can either lower our brows to protect our eyes, or raise our brows to increase our range of vision. Both will be helpful, but we have to choose. The brain has to assess which is the more important demand and instruct the face accordingly. If we look at monkeys, we see that during their most aggressive threats their brows are lowered; during their rather scared threats their brows are raised; and during moments of beaten submission their brows are lowered again. The position is much the same in man.

When a human being is very aggressive and might provoke an immediate retaliation, or where he is beaten and fearful of imminent attack, he sacrifices improved vision and protects his eyes with lowered eyebrows. When he is slightly aggressive but also very scared, or where he is in any state of conflict that does not appear to be in imminent danger of turning into a physical attack, he sacrifices eye protection for the tactical advantages of being able to see more clearly

The Eyebrow Cock expresses a contradictory condition of 'fearless fear'.

▼ There is one asymmetrical brow expression in which one eyebrow is lowered in a frown while the other is raised in surprise. Many people find this expression hard or even impossible to perform (as some do with that other asymmetrical expression the wink), but for those who use it regularly it is a valuable addition to the facial repertoire. Employed in contradictory moments it signifies a sceptical, quizzical frame of mind in which the performer is in a mood that is simultaneously forceful (and therefore frowning) and surprised (and therefore brow-raising). This condition of 'fearless fear' is visually expressed by means of the Eyebrow Cock.

what is happening around him, and raises his eyebrows.

Once these two actions have developed their primary roles, they can be used as indicators in quite mild contexts. We put them to work as contrived signals. A man can raise his eyebrows deliberately, even when he is not worried, simply in order to signal to someone 'How worrying for *you*'. But such refinements and modifications would not be possible were it not for the original primitive significance of the actions.

3. The Eyebrow Cock. This is a mixture of the previous two actions, with one eyebrow being lowered while the other is raised. It is not a particularly common expression and one which many people find hard to perform.

The message given by this action is as intermediate as the expression itself. Half the face looks aggressive while the other half looks scared. For some reason, this contradictory response is observed far more frequently in adult males than in females or juvenile males. The mood of the eyebrow cocker is usually one of scepticism. The single raised eyebrow acts rather like a question mark in relation to the other, glaring eye.

4. The Eyebrows Knit. The eyebrows are simultaneously raised and drawn towards each other. Like the last action, this is a complex one made up of two elements taken from the Lower and the Raise. The inward movement is taken from the Eyebrows Lower action, producing short vertical creases in the narrowed space between the eyebrows. The upward movement is taken from the Eyebrows Raise action, producing horizontal creases across the forehead. The knitting of the brows therefore produces a double set of skin wrinkles.

This is the expression associated with intense anxiety and grief. It is also observed in some cases of chronic, as opposed to acute, pain. A sudden sharp pain gives rise to the Face Wince reaction, with its lowered brows, but a dull prolonged pain is more likely to produce the Eyebrows Knit posture. It is the standard expression employed in headache advertisements.

In origin, this action appears to be an attempt by the brow to respond to a double signal from the brain. One message says 'raise the brows' and the other says 'lower them'. Different sets of muscles start to pull in opposite directions. The first set manages to pull the eyebrows up a little, but the second set, although trying to tug them down and in, only manages to pull them in towards each other.

In some cases, but by no means all, the inner ends of the eyebrows are pulled up further than the outer ends, resulting in the 'oblique eyebrows of grief'. This exaggerated form of the knitting action is most marked in professional mourners, who have seen more than their fair share of tragedy. If individuals with less tragic histories try to force their eyebrows up into the oblique position, they may have little success, even though they may be able to feel their brows trying to move into this special posture. Theoretically it should be possible to tell how much misfortune there has been in someone's past life simply by checking the ease with which they manage to adopt the oblique brows posture.

5. The Eyebrows Flash. The eyebrows are raised and lowered again in a fraction of a second. This brief, upward flick of the eyebrows is an important and apparently worldwide greeting signal of

▲ *A brief raising of the eyebrows is part of the complex* shrug *gesture. This action, employed generally as a disclaimer when expressing helplessness, innocence, blamelessness or irritation at someone else's stupidity, involves movements of the mouth, head, shoulders, arms and hands as well as the eyebrows when it is seen in its full form. Partial shrugs may, however, be limited to only one or a few of these parts of the body. In some instances, only the eyebrows move and this version is usually restricted to a particular social context. When two friends are sitting close together and a third party does something foolish, one friend may turn to the other and make a gestural criticism in the form of a brief raising of the eyebrows. This restrained form of shrugging has the advantage that it can be shared between the two friends without being obvious to the person who has acted foolishly.*

the human species. It has been recorded not only in Europeans from many areas, but also from as far afield as Bali, New Guinea and the Amazon basin, sometimes in instances where there have been no European influences. In each case it has the same meaning, indicating a friendly recognition of the other person's presence.

The Eyebrows Flash is usually done at a distance, at the beginning of an encounter, and is not part of the close proximity displays, such as hand-shaking, kissing and embracing, that follow. It is often accompanied by a head toss and a smile, but may also occur by itself.

In origin it is clearly a momentary adoption of the Eyebrows Raise posture of surprise. Combined with the smile it becomes a signal of pleasant surprise. The extreme brevity of the action, no more than a split second, indicates that the surprise mood is quickly banished, leaving the friendly smile to dominate the scene.

As already mentioned, the raising of the eyebrows has an element of fear in it, and it may seem strange that such a factor should play a part in a greeting between friends. But every greeting, no matter how amicable, involves an increase in social unpredictability. We simply do not know how the other person is going to behave or how he may have changed since we last met him. This inevitably tinges the encounter with a small, fleeting element of fear.

In addition to its role as a greeting signal, the Eyebrows Flash is frequently employed during ordinary conversational speech as a 'marking point' for emphasis. Each time a word is stressed strongly, the eyebrows flick up and back again. Most of us do this occasionally, but with some individuals it becomes particularly frequent and exaggerated. It is as if we are saying 'these are the surprise-points' in the verbal communication.

6. The Eyebrows Multi-flash. The Eyebrows Flash movement is repeated several times in quick succession. This is a joke gesture made famous by Groucho Marx and still widely employed by comedians in

■ Because female eyebrows are naturally less bushy than those of the male, women have often exaggerated this difference by plucking or shaving them and then re-defining them as slender pencilled lines. For males bushy eyebrows have always been acceptable as attractive masculine signals.

■ Where heavy male eyebrows meet in the middle, however, it is a different story. An old European rhyme issues the warning: 'Trust not a man whose eyebrows meet, For in his heart you'll find deceit.' A more common belief was that such a man was either a werewolf or a vampire. The origin of this superstition appears to have two elements. One is simply 'hairiness' – the covering of a part of the body (in this case the glabellum) with hair when it is not normally hirsute. This automatically links the owner to hairy fantasy figures such as devils and werewolves. The other factor is probably the permanent pseudo-frown which such eyebrows create. In an ordinary frown the glabellum is darkened by the added crease lines and we talk of a 'darkening brow' in this context. If there is dark hair growing over this patch of skin between the eyebrows, it gives the owner a sinister darkened brow the whole time and makes him seem permanently threatening and hostile. In such cases, even males may resort to a little secretive eyebrow-plucking from time to time. This would be necessary at fairly frequent intervals because eyebrows grow at about the same rate as scalp hair – about a third of a millimetre a day. The reason why they are always much shorter than scalp hairs is that they only last from three to five months before they fall out, whereas the hairs on top of the head survive for several years.

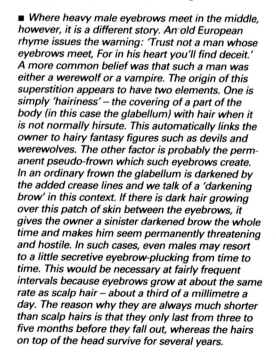

theatrical contexts. The eyebrows bob up and down quickly and repeatedly as a deliberately over-exaggerated, often mock-erotic, greeting signal. If the single Eyebrows Flash action says 'Hallo!', the joking Multi-flash in effect says 'Hallo-hallo-hallo!' If the single Flash says 'I am pleasantly surprised' by what I see, the Multi-flash says 'I am pleasantly astonished'.

7. The Eyebrows Shrug. The eyebrows are raised, momentarily held in the upper position and then lowered again. It is the brief 'hold' of the eyebrows in the raised position that distinguishes this action from the Eyebrows Flash of greeting and emphasis.

This is the eyebrow element of the complex shrug reaction, which also involves a special posture of the mouth, head, shoulders, arms and hands. Almost all the elements of this compound display can also occur separately, in isolation, or in conjunction with one or two other elements. The Eyebrows Shrug, although it may sometimes occur entirely on its own, is usually accompanied by a Mouth Shrug – a rapid and momentary turning-down of the mouth corners. This combination – what might be called the Face Shrug – frequently occurs in the absence of the other shrug elements.

Unlike the Eyebrows Flash, therefore, this action is typically linked with a 'sad' mouth rather than a 'happy' one. This should give it the meaning of a mildly unpleasant surprise, which is frequently the way it is used. If, for instance, two people are sitting together and a third

…a matter complicated by fashion and folklore.

■ *In recent years, the increased assertiveness of western females has led to a reduction in some of their more feminine fashions. Plucked eyebrows are less common and it is now possible for an actress who is considered beautiful to display strong, almost masculine brows. Here, the same actress is able to appear in both feminine and masculine roles without any alteration in her natural eyebrow shape. This would have been unthinkable a few decades ago.*

one nearby does something socially 'uncomfortable', one of the two intimates may give the Eyebrows Shrug to the other to indicate surprised disapproval.

The Eyebrows Shrug is also often seen as an accompaniment to speech in certain individuals. Nearly all of us, when talking animatedly, make repeated, small body movements to stress what we are saying. At each point of verbal emphasis we add a visual emphasis. With most people it is the hands or the head that keep moving with each stressed point, but some individuals favour their eyebrows. As they speak, their eyebrows are repeatedly shrugged in time with the special points they are making. This is typical of the speech of the chronic 'complainer', who seems to be perpetually surprised by the vagaries of life, but it is by no means confined to this particular personality type.

Leaving the question of eyebrow movements and turning to their anatomy, there is one important gender difference: male eyebrows are thicker and bushier than those of females. This difference has led to many forms of 'improvement'. Female eyebrows have been made super-female by artificially increasing their thinness and smallness. This has been done for centuries, using a variety of techniques such as shaving, plucking and painting. Originally the excuse given was that these procedures helped to ward off evil; later they were supposed to protect the body from disease and, in particular, to prevent blind-

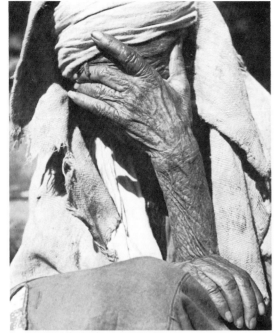

▲ Gestures involving hand-to-brow movements are quite common. Some are found worldwide, others are highly local. The most basic form of brow contact involves a simultaneous clasping of the head and covering of the face – a gesture of depression, defeat and dejection in which the person concerned does his best to cut off contact with the outside world. In ancient Greece and Rome the gesture of despair was formalized by beating the forehead with the fist. A more modest version of this persists today in the form of the palm clapped noisily to the brow at moments of shock and self-anger. It is the 'Oh no!' reaction of someone who has just realized he has done something extremely stupid. A more contrived form of this is the 'Shooting-oneself-in-the-temple' gesture in which a straight forefinger represents a gun. The 'Temple-screw' gesture, in which the forefinger rotates on the side of the forehead as if tightening up a screw on the brain-case, carries this meaning too, but can also refer to someone else having a 'screw loose'. All these gestures are based on the simple equation: brow = brain.

▲ The most drastic form of brow modification is the frontal crushing of the skull during infancy. Parents in certain tribal cultures impose this deformity on their infants while the bones of the skull are still soft, with the result that they grow up to be unusually flat-faced adults. Whether it causes any damage to the brain to be squashed in this way is not at all clear.

Hand-to-brow and brow-remoulding symbolism.

ness; later still, they were done to enhance personal beauty. In all instances the underlying urge was to make the eyebrows look more exaggeratedly feminine.

In recent times, the peak of eyebrow-plucking came in the inter-war period, in the 1920s and 1930s, when 'the eyebrow pencil was in every vanity bag, available in five bewitching shades.' After the thickness of the eyebrows had been reduced by the use of tweezers, the pencils were employed to emphasize the thin arched line of hairs that survived. If a particular woman felt that her eyebrows were in an unbecoming position on her forehead, she could, of course, remove them altogether and paint in totally artificial lines in a new position to suit her taste. When this step was taken, the new eyebrows nearly always appeared *above* the true position, giving the face less of a 'frown'.

Perhaps the strangest example of false eyebrows comes from England in the early eighteenth century. At this time, fashionable eyebrows were also shaved and then replaced, but it was the nature of the replacement that was so bizarre. The eyebrow 'falsies' of the day were made of mouse-skin. Swift records this strange fashion with the words: 'Her eyebrows from a mouse's hide, Stuck on with art on either side.'

Because these attentions were all concerned with improving a lady's looks, it followed that leaving the eyebrows in their natural unplucked condition was frequently a statement of a strong 'non-sexual' nature. Women working in conditions where they were expected to suppress their sexuality were also expected to leave their eyebrows alone. In the 1930s there was a hotly debated case involving a London hospital where a matron had refused permission for a nurse to pluck her eyebrows. A complaint was made to the effect that this was interfering with personal liberty, but the matron's decision was upheld by the London County Council. Hospital patients were, as a result, 'spared' the erotic stimulation of the sight of a delicately plucked eyebrow, as they lay in their sick-beds.

Apart from eyebrow-shaving and -shunting, the main alteration to the brow region has been concerned with the remoulding of the forehead of tiny infants. Several cultures in quite different parts of the world have practised head-flattening of one kind or another. The ruling families of Ancient Egypt sometimes followed this fashion and it has also been discovered in skulls from thirteenth century Sweden, from the provincial regions of France as late as the nineteenth century, from several groups of American Indians, and from Nazi Germany in the 1930s. The reasons have varied. Sometimes a flattened forehead was simply considered more beautiful. Sometimes it was said to improve the intellectual capacity by reshaping the brain. Sometimes it was to demonstrate high social status, because the brow-flattened individual was unable to perform the menial task of carrying water-pitchers on top of the head. In Germany during the National Socialist regime the warped Nazi theories about Aryan purity drove some parents to the extreme of attempting to improve the cranial index of their offspring by binding their skulls. As any parent knows, a baby's skull is alarmingly soft during the early days of life and it is easier than one might imagine to reshape it during this first, tender stage.

THE EYES

■ The eyes are the dominant sense organs of the human body. It has been estimated that 80 per cent of our information about the outside world enters through these remarkable structures. Despite all the talking and listening we do we remain essentially a visual animal. In this we do not differ greatly from our close relatives, the monkeys and apes. The whole primate order is a vision-dominated group, with the two eyes brought to the front of the head, providing a binocular view of the world.

The human eye is only about an inch in diameter and yet it makes the most sophisticated television camera look like something from the Stone Age. The light-sensitive retina at the back of the eyeball contains 137 million cells which send messages to the brain, telling it what we are seeing. Of these, 130 million are rod-shaped and concerned with black-and-white vision; the remaining 7 million are cone-shaped and facilitate colour vision. At any one moment these light-responsive cells can deal with one-and-a-half million simultaneous messages. Because it is so complex, it is hardly surprising that the eye is the part of the body which shows the least growth between birth and adulthood. Even the brain grows more than the eye.

At the eye's centre is the black spot we call the pupil – the aperture through which light passes to fall on the retina. The pupil increases in size with weak light and decreases in strong light, controlling the amount of illumination falling on the retina. In this respect the eye does act much like a camera with an adjustable diaphragm, but it also has a curious override system. If the eye sees something it likes very much the pupil expands rather more than normal, and if it sees something distasteful it shrinks to a pinprick. It is easy to understand the second of these two responses because further contraction of the pupil's aperture would simply reduce the illumination of the retina and 'damp down' the distasteful image. The increased pupil dilation that occurs when we see something attractive is harder to explain. This must interfere with the accuracy of our vision by letting too much light flood on to the retina. The result must be a hazy glow rather than a sharp balanced image. This may be an advantage for young lovers, however, when they gaze deeply into each other's dilated pupils. They may benefit by seeing a slightly fuzzy image bathed in a halo of light – the very opposite of a 'warts-and-all' image.

Around the pupil is the muscular, coloured iris, the contracting disc which is responsible for the changes in pupil size. This task is performed by involuntary muscles, so we can never deliberately or consciously control our pupil size. It is this fact that makes pupil expansion and contraction such a reliable guide to our emotional responses to visual images. Our pupils cannot lie.

The colour of the iris varies considerably from person to person, but this is not due to a variety of pigments. Blue-eyed people do not possess blue eye pigment; they simply have *less* pigment than others, and this gives the impression of blueness. If you have a dark brown ring around your pupils, this means that you have a generous amount of melanin pigment in the front layers of your irises. If the amount of melanin here is less and the pigment is largely confined to the deeper layers of the iris, then your eyes will be paler, ranging from hazel or green to grey or blue as the pigment decreases. Violet colouring is due to blood showing through. Brightly coloured eyes in

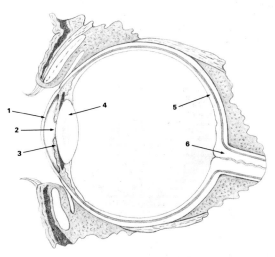

■ The human eye is the most extraordinary organ in the human body. It is capable of responding to one-and-a-half-million simultaneous messages, and yet it is no bigger than a table-tennis ball. As with other primates, our flattened faces enable us to enjoy binocular vision, with accurate depth assessment. Evolved originally in connexion with a leaping lifestyle in the treetops, it served us well during our primeval hunting days, when estimating distances was of vital importance. Key features of the eyeball include: (1) the cornea; (2) the pupil; (3) the iris; (4) the lens; (5) the retina; and (6) the optic nerve.

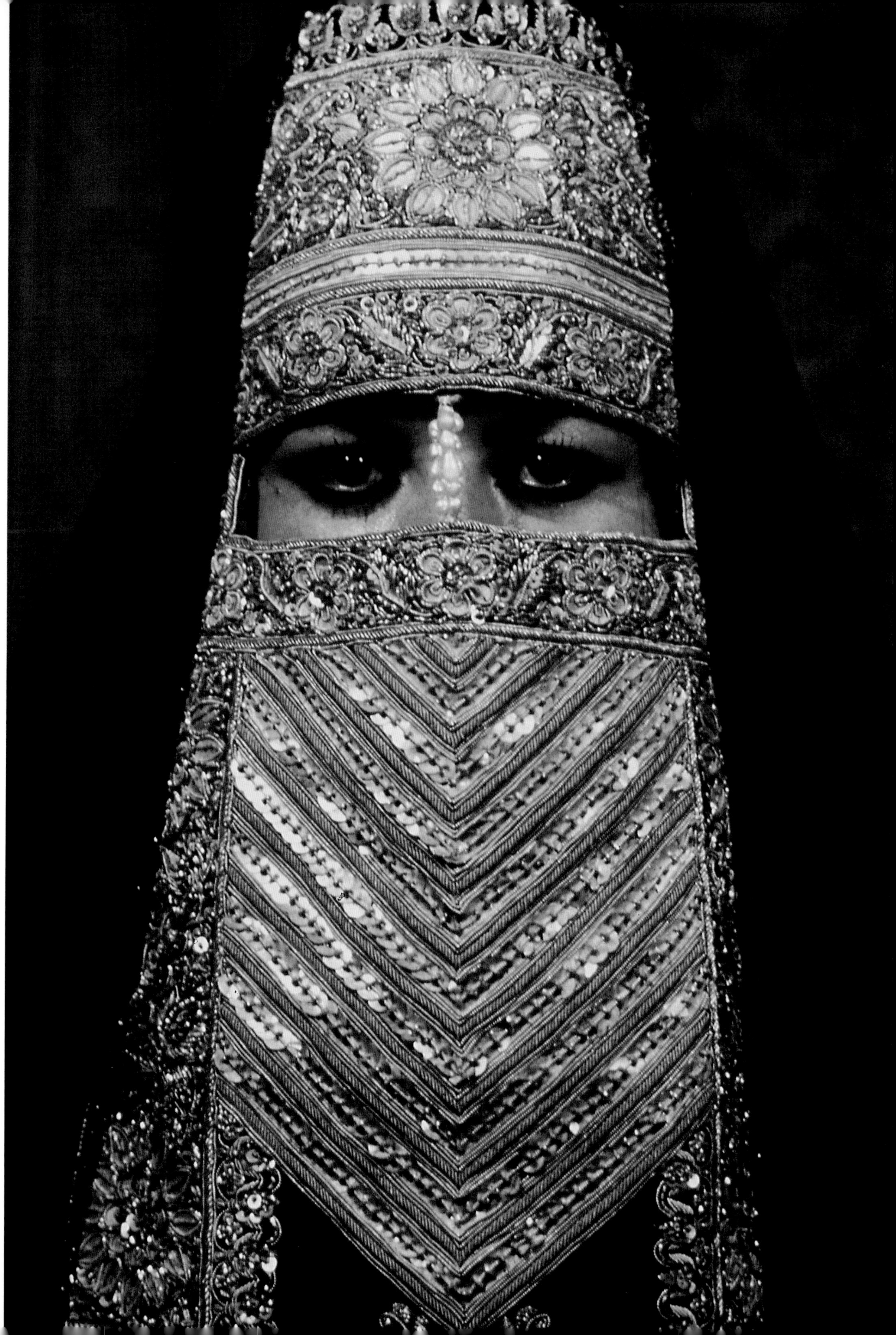

Human eyes, unlike those of monkeys and apes, have whites to signal gaze direction.

51

■ Elaborate costumes frequently cover other sense organs such as the mouth, the nose and the ears; but even where the aim is to mask the entire body the eyes must remain free to communicate with the outside world.

■ Unlike other species of primates, the human being has visible whites to the eyes. In most monkeys and apes the corresponding parts of the eyes are dark brown and this makes it less easy to read the exact direction of their gaze. But in humans, where it has become vital to follow the shifting attention of our social companions, the conspicuous white patches on either side of the coloured iris tell us immediately the angle of the glance.

We all have 'slant eyes' in the womb; only Orientals retain them after birth.

▼ *The so-called 'slant eyes' of Orientals are the result of the presence of a flap of skin called the 'epicanthic fold'. This is an infantile characteristic which we all possess when we are in the womb, but which westerners lose before they are born. It seems that the Oriental race evolved originally in extremely cold regions of the world, where the eyes needed extra protection against freezing conditions.*

human beings are therefore something of an optical illusion. They indicate a loss of melanin and seem to be part of the general 'paling' of the body that occurs as one moves away from the equator towards the less sunny polar zones. This effect is most striking when one compares the babies of white people with those of darker skinned races. Almost all white babies are blue-eyed at birth. Dark-skinned babies are dark-eyed. Then, as they grow older, most white offspring gradually develop the melanin pigment on the front of the iris, making their eyes darker and darker. Only a very small percentage fail to do this and retain their 'baby blues'.

Covering the pupil and the iris is a transparent window, the cornea, and around it is the region we refer to as 'the whites of the eyes', technically called the sclera. It is this non-optical part of the human eye that is its most unusual feature. Uniquely, parts of the whites of our eyes are visible to onlookers. Most animals have circular 'button' eyes. The same is true of lower primates, but with most monkeys the skin around the eyes is slightly pulled back, left and right, to give the eyes 'corners'. These eyes are still nearer to a circle in shape than an oval; but the trend goes a step further in the apes, where the eyes are more elliptical, approaching the human shape. Even here no 'whites' are visible, the exposed area on either side of the iris matching its dark brown colour. In humans the whiteness of these same areas makes them highly conspicuous. The effect of this small evolutionary change is that during social encounters shifts in gaze direction are easily detected, even at a distance.

Surrounding the visible part of the eye, the eyelids, fringed with curved eyelashes, have greasy, shiny edges. The greasiness is caused by the secretions of rows of tiny glands, visible as minute pinpricks just behind the roots of the eyelashes. The regular blinking of these lids moistens and cleans the cornea. The process is aided by the secretion of tears from the tear gland, tucked in under the upper eyelid. The liquid is drained off through two small tear ducts – also visible as pinpricks, but bigger ones, on the edges of the eyelids. They are positioned at the nose-end of the lids, one in the upper and one in the lower. The two ducts connect up into a single tube which carries the 'used' tears down into the interior of the nose and away. When emotions or an irritation in the eye makes the tear glands produce tears more quickly than the ducts can drain them off, we weep. The excess tears spill over on to our cheeks and we wipe them away. This is the second unique feature of the human eyes, for we are the only land animal that weeps with emotion.

Between the two tear ducts, in the corner of the eye next to the nose, there is a small pink lump. This is the remnant of our third eyelid and it now appears to be completely functionless. In many species it is an organ of some value. Some use it as a 'windscreen wiper', blinking it sideways to clean the eye; others have coloured ones which they flash as a signal; still others have completely transparent ones they can use as natural sunglasses. Diving ducks go even further, having specially thickened transparent ones which they pull over their sensitive corneas when swimming under water. If only our primeval ancestors had been more aquatic our sub-aqua pleasures today might have been greatly enhanced.

The eyelashes, which provide us with a protective fringe above and

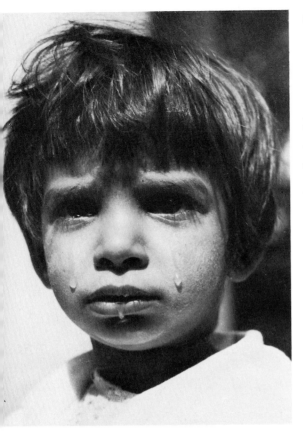

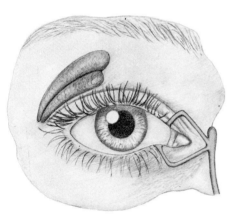

▲ *The human species is the only land animal that weeps copiously with emotion. The normal secretion of tears is connected with cleaning the corneal surface of the eye, but in moments of intense passion and excitement abnormally powerful secretion occurs, so that the tears spill out and down the cheeks. This may have evolved simply as a visual display of distress, or it may be a cunning way of removing excess stress chemicals from the system. The tear fluid is secreted by a gland which lies just above the eye and drained by a pair of ducts near the eye's inner corner.*

below the eyes, have one exceptional feature: they do not become white with age like other head and body hairs. Each eye has about 200 of them, more on the upper lid than the lower, and each lash lasts between three and five months before falling out and being replaced. They have the same lifespan as the hairs of the eyebrows.

Another form of eye protection occurs in Orientals, who possess a flap of skin called the epicanthic fold which lies over the upper eyelid and gives Mongoloid eyes their characteristic 'slant'. This fold is present in the human foetus in all races, but is only retained into adulthood by the eastern branch of the human family. A few western babies are born with the eye fold still present, but it gradually disappears as the nose narrows and changes shape with advancing age. Among the Oriental peoples the epicanthic fold seems to have been retained as part of a general adaptation to cold. The whole face is more fat-laden, flatter and better able to cope with icy conditions, and the extra fold of skin above the eyes helps to shield this delicate area in an extreme environment.

There are small gender differences in the eye. The male eye is very slightly bigger than the female, while the female eye shows a higher proportion of white than the male. In many cultures the tear glands are more active in emotional females than in emotional males, but whether this is due to cultural training which requires the males to be less emotionally demonstrative or is a biological difference of a more basic kind is hard to say. It does appear to be a remarkably widespread difference for it to be simply the result of social training.

A word about the tears themselves: they are not only lubricants for the exposed surface of the eye, they are also bactericidal. They contain an enzyme called *lysozyme* which kills bacteria and protects the eye from infection.

Temporary eye-strain is a common affliction of civilized man. The fact is that our eyes evolved to work efficiently at much longer distances than are usually encountered in modern life. Prehistoric men did not sit bent over desks or slumped in armchairs, poring over figures, reading small print, or watching flickering images on screens. As they were hunters their eyes were more concerned with images in the far distance. More effort is needed for the eye muscles to focus the eyes on a near object than a far one, so the close-vision urbanite can easily give himself muscle fatigue by spending hour after hour peering at a spot only a few feet in front of him. When we watch television or read a book it is not just proximity that causes the problem but lack of variation in the depth of vision. This forces the eye muscles to hold a particular degree of contraction for an unnaturally long period. Our eyes may ache, but that does not mean we have damaged them any more than a man who runs a mile has damaged his aching legs. All they need is a rest. The solution is simply to look away from the screen or page occasionally and focus on some more distant object for a few moments.

Poor eyesight must have been a curse for many of our remote ancestors, not only because of the lack of precision in obtaining visual information but also because the permanent strain of trying to see with defective vision causes severe headaches and migraines. The curse remained for people of the earliest civilizations and with the invention of writing it became acute, many elderly sages and scholars

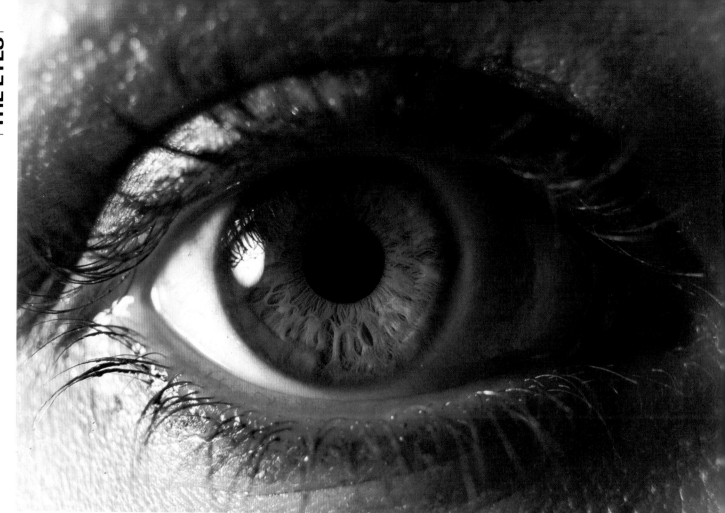

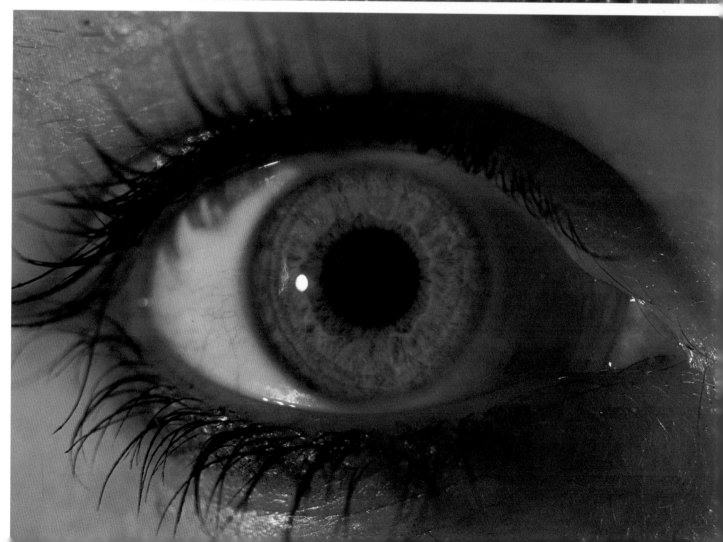

■ *The pupils of the eyes vary dramatically in size according to the amount of light falling on them, and they also dilate and contract under emotional influences. If we see something we like, the pupils grow bigger; if we see something distasteful, they shrink to pinpricks. The muscular iris which surrounds the pupil varies in colour, according to the amount of melanin present. White babies are nearly all blue-eyed, but most of them develop darker eyes as they grow older. Dark-skinned babies are dark-eyed from birth. There is no blue pigment in blue eyes or green pigment in green eyes. These and other subtle greys and violets are the result of reduced amounts of melanin in different layers of the iris. It is a myth that blue eyes are more delicate than dark ones. They may be more sensitive to light but they are just as strong in other respects.*

having to employ young men to read for them. Seneca, the Roman connoisseur of the art of rhetoric who lived at the time of Christ, seems to have been the first person to attempt to solve this terrible problem. It is said that despite poor sight he managed to read his way through the libraries of Rome by using a 'globe of water' as a magnifying glass. This ingenious solution should have led to an early development of eyeglasses, but it failed to do so. It was not until the thirteenth century that the English philosopher Roger Bacon recorded his observation that 'If anyone examines letters or other minute objects through the medium of crystal or glass...if it be shaped like the lesser segment of a sphere, with the convex side towards the eye, he will see the letters far better and they will seem large to him.' He went on to say that such a glass would be useful to those with weak eyes, but again there was no rush to develop this boon to human vision. Towards the end of the century, in Italy, true spectacles for reading did at last appear, though it is not clear whether they were influenced by Bacon. In 1306 a monk in Florence gave a sermon which included the following phrase: 'It is not yet twenty years since the art of making spectacles, one of the most useful arts on earth, was discovered...' At about the same time Marco Polo recorded seeing elderly Chinese using lenses for reading, so it is clear that by the fourteenth century the move towards a wide use of eyeglasses had begun in earnest. In the fifteenth century special lenses for correcting short-sightedness

Framing the eyes with the hands transforms the gaze into a super-stare.

appeared, and in the eighteenth century Benjamin Franklin invented bifocals. The first successful contact lenses were made in Switzerland in 1887.

This brief history of eyeglasses is of more than medical interest because it also changed the appearance of our eyes. The shape of the spectacles became part of the facial expression of the wearer. A heavy upper rim became a super-frown, making the owner look more fierce and domineering. A wide circular rim produced a wide-eyed stare, as though the curve of the rim represented arched eyebrows. There was no deception, as in the case of subtle make-up. The spectacles were clearly not part of the face and yet it was impossible not to be influenced by their lines, just as an eye-mask alters the whole expression of the wearer.

The effect of dark glasses is especially dramatic. Tell-tale eye movements, made conspicuous by the whites of the eyes as mentioned earlier, provide a constant source of information during social encounters, but dark glasses effectively eliminate that information. Darting eyes, shifty eyes, inattentive eyes, over-attentive eyes, dilated eyes – all are hidden from the companions of the man wearing 'shades'. They can only guess at what is taking place behind the mask of his sunglasses.

What are they missing? Precisely what do human eye movements tell us? In any social gathering subordinates tend to look at dominant figures and dominants tend to ignore subordinates, except in special circumstances. For example, if a friendly submissive individual enters a room his eyes will dart this way and that, checking on all those present. If he spots a high-status, dominant individual he will keep a persistent watchful eye on him. Whenever a joking remark or a controversial statement is made or a personal opinion is expressed the subordinate's eyes will flick in the direction of the dominant person to assess his reaction. The 'boss' figure typically remains aloof during such exchanges and hardly bothers to look at his subordinates during general conversation. But if he fires a straight question at one of them he does so with a direct stare. The individual on whom he fixes finds himself unable to return this stare for any length of time and during most of his reply looks elsewhere.

This is the situation where a clear pecking order operates and where certain individuals have control over others and wish to exercise it. When friends of equal status meet, the eye movements are rather different. Here everyone uses 'subordinate' eye movements even though they are not subordinate. This is done because the simplest way of demonstrating friendliness with body language is to display non-hostility and non-dominance. So we are watchful of our friends, treating them with our eyes as if they are dominants. When they speak or are active we look at them; when we speak and they watch us we look away and glance at them only briefly from time to time to check their reactions to what we are saying. In this way, each of two friends will treat the other as the powerful one and thus make each other feel good.

If a dominant individual wishes to ingratiate himself with someone he can do so by deliberately adopting the friendly body language of an equal. When addressing an employee or servant of some kind, he can manipulatively switch on an attentive eye, hanging on the

■ *The 'spectacles' or 'binoculars' gesture, which is usually a playful way of saying 'I can see you', has the effect of transforming the face, making it look as though it is indulging in a super-stare. This gives the gesturer a slightly threatening demeanour, even when no threat is intended. Heavy-rimmed spectacles operate on the same basis, providing the wearer with a permanently 'menacing', artificial gaze. This is because, like all primates, we are programmed to respond with a degree of apprehension to a pair of widely staring eyes.*

underling's every word. Such devices are rarely used by dominant individuals outside special contexts such as election campaigns.

Prolonged eye-to-eye staring occurs only at moments of intense love or hate. For most people in most settings a direct stare that is held for more than a few moments is too threatening and they quickly look away. For lovers there is such total mutual trust that they can hold each other's gaze without even a twinge of fear. As they stare into each other's eyes they are unconsciously checking the degree of pupil dilation. If they see deep black pools they know intuitively that their feelings are reciprocated. If they see tiny pinprick pupils they may start to feel uneasy, sensing that all is not well in their relationship.

Turning from lovers to haters, the staring eyes of an angry person are strongly intimidating. In earlier days when superstitions were rife it was believed that supernatural beings were watching over human events and influencing their outcome. The fact that these divine powers or deities were *watching* meant that they must have eyes. Since they had to watch over so much it was supposed that they must have many eyes and be all-seeing. Where good gods were concerned this was clearly a great advantage to human beings because benign deities could be protective. But there were also bad gods, demons and devils – evil spirits with evil eyes – and a look from them could spell disaster.

A belief in the power of evil eyes became widespread and survives even today in some parts of the world, such as southern Italy. The evil eyes became the Evil Eye, a malicious, harmful and even deadly influence which could strike down a victim without warning. If its glance fell on you something terrible would happen. Sometimes an ordinary person became possessed, against his will, of the Evil Eye, and everyone he looked at suffered in some way soon afterwards. At least two Popes – Pius IX and Leo XIII – were possessed of this terrible affliction, which caused nightmare problems for their faithful followers.

Judges trying cases of possession by the Evil Eye during the Inquisition insisted on prisoners being led into court backwards so that

Spectacles are clearly not part of the face.

their lethal glance could not cause havoc in the courtroom. Many people in widely different cultures wore special talismans and amulets to help protect them. Some of these were obscene, the idea being that they would distract the Evil Eye and divert its gaze. Others consisted of an image of a staring eye to 'out-stare' the Evil Eye. Some babies were even tattooed with a protective eye on their backs in case the Evil Eye sneaked up from behind.

Boats, houses, necklaces, idols, vessels, shops – all were decorated with protective eyes, eyes that were fixed and never blinked and could never be out-stared. Though we may laugh at these primitive super-

■ *The signalling effect of glasses has often been exploited. The domineering personality emphasizes his masterful style with heavy, frowning spectacles; the shy pop star hides his modesty behind eccentric 'shades'; the peace-campaigning celebrity hopes to demonstrate his lack of aggression, and the quiet man his lack of affectation, by selecting 'granny-style' spectacles with wire-thin circular rims. One of the more popular devices is the wearing of very dark glasses even when there is no sunshine. This is the strategy of the smooth operator, the cool negotiator, the shady dealer, the incognito star – and the nonentity hoping to be mistaken for the latter. For some, dark shades offer a simple means of disguising identity, but for others it is a trick to avoid revealing telltale eye movements.*

Yet it is impossible not to be influenced by them.

stitions we are still not free of them. The belief in the threatening power of the staring eye remains with us in several less-than-obvious ways. The lucky horseshoe is one of them. Originally it symbolized the female genitals, and placing it over a door for good luck was a device for distracting the Evil Eye from staring into the building.

Since the Evil Eye's worst deeds were thought to be caused by envy it was important not to lavish praise on someone who might be vulnerable. To call someone wonderful at moments of high risk was considered unforgivable in some quarters. It was tempting disaster, encouraging the Evil Eye to strike down the 'wonderful one' out of uncontrollable envy. In equine circles the owners of horses were fearful that praise for a particular animal might lead to the Evil Eye breaking the creature's leg in a fit of pique. Instead of uttering words of praise they therefore themselves said 'break a leg' so that the Evil

The power of the gaze to express love and hate, to elevate and degrade, has always impressed.

Eye would then feel superfluous and leave the animal alone. This curious form of covert 'good wishes' survives today in that other high-risk arena the theatre, where it is said to an actor before he makes his entrance.

All of these 'good luck' techniques originally came about because of the false belief that a stream of energy similar to the fire of the sun was emitted by the eye whenever it looked at something. Even Leonardo da Vinci thought that there was a beam of light streaming from the open eyes and that it actually made contact with the objects the eyes looked at. Clearly if eyes were supernaturally powerful and also malevolent the stream of light energy pouring from them on to a victim could easily be harmful and perhaps lethal. Once it was realized that light fell on to the eye rather than emanated from it, the basis for the Evil Eye superstition was destroyed. This makes it even more surprising that it has lingered on so stubbornly around the shores of the Mediterranean and elsewhere.

Things that do emanate from ordinary eyes today are the various eye expressions that transmit visual signals to onlookers, telling them about the changing moods of their companions. Most of these expressions are too well known and obvious to warrant discussion, but some of them deserve a brief comment.

Lowering the eyes formally is sometimes used as a modesty signal. It is based on the natural behaviour of subordinates who dare not look at their superiors, but it is not random in its direction. The primly modest 'flower' does not cast her eyes to left or right but only down at the ground. There is the suggestion of a bow in this action, or of the lowering of the head in submission.

Raising the eyes is also sometimes used as a deliberate signal. If they are held in the up position for a while, the expression is one of 'a pretence of innocence'. Performed today only in jest, this eye movement is based on the idea of looking up to heaven as a witness of the claimed innocence.

Glaring eyes are often employed by parents when trying to subdue children while remaining silent. The glare is a complex version of the stare. The eyes fixate the 'victim' with frowning eyebrows but widely opened eyes. This is a contradiction because the opening wide of the eyes normally goes with raised eyebrows, so these two parts of the face have to work against each other. For this reason it is not an expression that is held for any length of time. During the glare the upper eyelids press upwards so hard that they almost disappear behind the descending brows, and the demarcation line of the glaring eye is provided by the brow-skin, not the eyelid. This gives a strange apparent shape to the eyes which is unmistakable. The message of the glare is one of surprised anger.

The sidelong glance is used to steal a look at someone without being seen to do so. It is also used as a deliberate signal of shyness, when it becomes a sign of coyness. 'I am too frightened to look straight at you, but I can't help staring' is the message here, and the popular phrase 'making sheep's eyes at someone' has been coined to describe this action.

De-focusing the eyes occurs when we are very tired or when we are daydreaming. Someone who wishes to signal that they have something special to daydream about (a new lover, for instance) may delib-

■ *There are only two emotional conditions under which two people will stare closely into one another's eyes for any length of time. The first is intense love and the second is intense hate. Lovers gaze deeply into one another's eyes unconsciously checking their partners' pupil dilation, which reveal their true feelings. The greater the dilation the greater the love. Haters stare at one another to try and make the rival look away, which would count as a psychological defeat.*

▶ *People who believe in the importance of good and bad luck in their lives are often superstitious enough to decorate their rooms with eyes. These are known technically as 'apotropaic eyes' and have been used in one style or another for thousands of years. The idea behind them is that if the 'evil eye' comes to look upon the person in the room and brings bad luck with its glance, it can be out-stared by the other eyes present. Because they are un-blinking, artificial eyes, they will eventually be able to force the 'evil eye' to look away, and no harm will then come to the owner of the room. Similar eyes have been fixed onto the prows of boats since ancient times and can still be seen around the shores of the Mediterranean to this day.*

erately stare out of a window or across a room with de-focused eyes as a way of impressing companions.

Widening the eyes to the extent that white is showing above and/or below the iris is a basic response of moderate surprise. This action increases the field of vision of the eyes and paves the way for increased responsiveness to visual stimuli. As with many of these automatic reactions of the eyes, we now use a deliberate 'acted' version to signal mock-surprise.

Narrowing the eyes also has its deliberate version. Basically it is a protective response against too much light or possible damage, but it has a contemptuous form in which the person narrowing the eyes is plainly *not* suffering overexposure or a threat of damage. This artificially 'pained' expression implies that those present are the cause of a more-or-less permanent anguish. It is an expression of distaste – a haughty look of disdain upon the world around. The special fold of skin above the oriental eye sometimes creates a false impression of 'haughtiness' because it makes the eye look as though it is being deliberately narrowed.

Glistening eyes transmit an entirely different signal and one that is hard to fake (except for professional actors). The twinkling, sparkling or glistening surface of the eyes is slightly overloaded with secretion from the tear glands caused by aroused emotions, but the feelings are not sufficiently strong to produce actual weeping. These are the gleaming eyes of the passionate lover, the adoring fan, the proud parent and the triumphant athlete. They are also the shining eyes of anguish, distress and bereavement, in fact any strong emotional condition that stops just short of crying.

Weeping itself is also a powerful social signal. The fact that we weep and other primates do not has aroused considerable interest, and it has been suggested that this difference is due to our ancestors having passed through an aquatic phase several million years ago. Seals weep when emotionally distressed but land mammals do not – except for man. Sea otters have also been seen to weep when they have lost their young, and it is suggested that copious shedding of tears is a by-product of the improved eye-cleaning function of tears in mammals that have returned to the sea.

This aquatic explanation is certainly logical enough. If man went

The eloquence of the eyes is unrivalled.

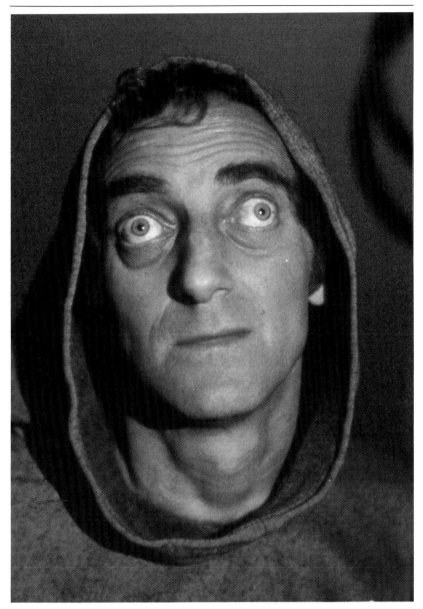

■ *These are five of the many different eye expressions with which we signal our feelings to our companion. Basically they are related to the degree to which we wish to close our eyes to protect them, or open them wider than usual in order to be able to take in the broadest view of our surroundings. The narrowed eyes of the outdoorsman are a simple response to too much sunlight, but a similar expression can be adopted by someone in dim interior light who wishes to transmit a signal of pained distaste. The wide eyes of surprise can be artificially exaggerated by the application of certain kinds of make-up, a device popular with Indian dancers, whose eye movements are an integral part of their performance. Sometimes one eye stays open while the other closes, an action we refer to as winking. This is really directional eye-closing, in which we aim the closed eye at a friend, implying that there is something hidden – a secret we share. If performed to a stranger of the opposite sex it implies that there is shared (and as yet unspoken) sexual desire. The mock-innocent eyes of the comedian who gazes upwards have a double meaning. They pretend shyness by looking away from the fondled girl, and they are also aimed up to the heavens as if seeking God's forgiveness.*

through an aquatic phase several million years ago, stepped up his tear production in response to prolonged exposure to sea water and then returned to dry land as a savannah hunter, he might well retain his tearful eyes, exploiting emotional weeping as a new social signal. It would explain why he is the only primate to display this characteristic. An alternative explanation is that it was the dusty world of the savannah which increased tear production and that copious emotional weeping was a by-product of improved eye-cleaning. If it is pointed out that other mammals living in dusty conditions are all non-weepers when distressed, it can be argued that they all have hairy cheeks in which flowing tears would be lost. Only on the naked facial skin of the human species would the glistening teardrops act as a powerful visual signal to nearby companions.

A completely different explanation of weeping eyes is based on the idea that tears, like urine, have excretion of waste products as their main function. Chemical analysis of tears produced by distress and

■ *All the various eye gestures are made conspicuous by the fact that we tend to concentrate on the eye region when we meet at social gatherings. A twitching foot is far less likely to be noticed than a twitching eye. Even so, we are not content to let matters rest there and, for thousands of years, have resorted to different ways of making our eyes more conspicuous. Since the earliest civilizations, people have darkened the eyelashes to make their movements more obvious, have outlined the eyelids to exaggerate the shape of the eyes, and have painted the skin around them to heighten the contrast of the face-colour with the whites of the eyes. Today this type of make-up is almost exclusively confined to females, but this has not always been the case. Back in the seventeenth century, John Bulwer reported that 'The Turks have a black powder...which with a fine pencil they lay under their Eye-lids.' The illustration supplied by his artist depicts, not a female Turkish beauty, but a heavily mustachioed male.*

The People of *Candou* Ifland put a certaine blackneffe upon their Eye-lids.

The *Turks* have a black powder made of a Mineral called *Alchole,* which with a fine pencill they lay under their Eye-lids, which doth color them black, whereby the white of the Eye is fet off more white : with the fame powder alfo they colour the haires of their Eye-lids, which is practifed alfo by the Women. And you fhall finde in *Xeno-phon,* that the *Medes* ufed to paint their Eyes.

The cosmetic enhancement of eye impact.

those produced by irritation to the surface of the eyes has revealed that the two liquids spilling down the face contain different proteins. The suggestion is that weeping is primarily a way of ridding the body of excess stress chemicals, which would explain why 'a good weep makes you feel better' – the improvement in mood being a biochemical one. The visual signal of the wet cheeks of the weeper, which encourages companions to embrace and comfort the distressed individual, must then be seen as a secondary exploitation of this waste-product-removal mechanism. Once again it is hard to see how this theory can be reconciled with the absence of weeping in such animals as chimpanzees, who suffer from intensely stressful moments during social disputes in the wild.

Leaving the dramatic subject of weeping and turning to the more mundane one of blinking, there are several deliberate signals which are in use today. The ordinary blink, the windscreen-wiper action of the eyelids which cleans and moistens the corneal surface at frequent intervals througout the day, takes approximately ¼40th of a second. In emotional states, as tear production starts to increase, the blink-rate increases with it, so that a measure of the frequency of blinking can be used as an index of mood.

Modified forms of blinking include the Multiblink, the Superblink, the Eyelash Flutter and the Wink. The Multiblink occurs when someone is on the verge of tears. It is a desperate attempt to bail out the eyes before they start spilling over. Because of this it can also be used as a conscious signal of sympathetic distress. The Superblink is a single massive exaggerated blink, slower in speed and greater in amplitude than the normal blink. It is used as a melodramatic signal of mock-surprise and is employed exclusively as a contrived 'theatrical' action. The message is 'I don't believe my eyes, so I am wiping them clean with a huge blink in order to make sure that what I am seeing is really there.' The Eyelash Flutter in which the eyes are rapidly fluttered open and shut is similar to the Multiblink but involves a greater degree of eye-opening, being performed with a wide-eyed 'innocent' look. It is another contrived coquettish action of a theatrical kind, employed in a 'you can't be cross with little me' context.

The wink is a deliberate one-eyed blink that signifies a state of collusion between the winker and the person winked at. The message of the wink is 'You and I are momentarily involved in a shared act which secretly excludes all others.' Performed between friends at a social gathering it implies that the winker and his companion are privately in sympathy over some issue, or that they are closer to each other than either is to the others present. Performed between strangers the gesture usually carries a strong sexual invitation regardless of the genders involved. Because it suggests a private understanding between two people the secretive wink can be used openly as a 'tease' gesture to make a third party feel like an outsider. Whether it is used covertly or overtly the gesture is considered improper by writers on etiquette, one authority declaring that 'in Europe the act of winking by a woman is not "upper class" and might be classified with the jab in the ribs to make a point...' In origin the wink could be described as a 'directional eye closure'. Closing the eye suggests that the secret is aimed only at the person being looked at. The other eye is kept open for the rest of the world, who are excluded from the private exchange.

THE NOSE

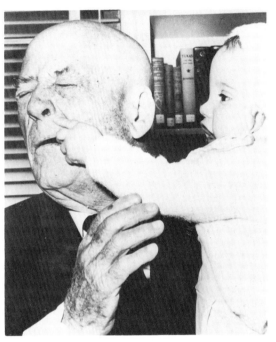

■ The human nose, with its prominent bridge, its elongated tip and its downturned nostrils, is unique. Our nearest living relatives, the monkeys and apes, have nothing quite like it. Those that do have a long snout also have a long face to match. We have a long snout on a flat face, and this strange condition requires a special explanation.

Some anatomists have offered the lame argument that, as the human face became flatter during the course of evolution, the nose simply stayed where it was, like a large rock being exposed by the receding tide. It is hard to accept this view. There is something so positive about the independence of the nose from its surrounding facial elements, that the 'projectile organ' as it has been called, must give some specific biological advantage to its owners. Several have been proposed.

The first theory sees the proud human proboscis as a resonator. Its enlarged condition is interpreted as a move in support of the ever-growing importance of human vocalization. As the voice evolved and speech developed, so, it is argued, did the nose. To illustrate this it is only necessary to try talking while pinching the nose shut between thumb and forefinger. The loss of vocal quality is dramatic. This is why opera singers are so terrified of catching a cold. But perhaps the clear human voice needs only the large sinuses – the hidden nasal cavities – to resonate efficiently? In which case we still need some other explanation for the protruding, external nose.

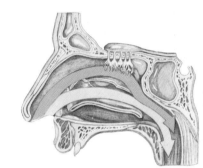

A second theory looks at the nose as a shield – a bony piece of armour helping to protect the eyes. If you place the tip of your thumb on your cheek-bone, a fingertip on your eyebrow and another on the bridge of your nose, you will feel your hand pushing against three defensive protrusions surrounding the eye. This bony triangle protects the soft and vulnerable eye from frontal blows. Such protection must have become more vital when our ancient ancestors switched from a gentle, fruit-picking existence to the risks and dangers of the hunt. Supporting this theory is the battered face of any veteran boxer: the bridge of the nose reveals the punishment it has received, while the eyes are usually still sharp and alert.

■ The human nose, besides sniffing strange odours (above), acts as a vital air-conditioning unit, warming, cleaning and moistening the air we breathe in, before it reaches the delicate lungs. Assisting in this – and also adding resonance to the voice – are the nasal sinuses (below), but the price we pay for possessing these valuable cavities is an all too common susceptibility to local infections.

A third and rather fanciful idea sees the nose as a shield of a different kind – a shield against water. It is argued that our ancestors may have gone through an aquatic phase several million years ago and that during those watery days our bodies adapted in a number of ways. In this view, the nose is seen as a shield against the inrush of water when diving. It is pointed out that when we *jump* into the water we hold our noses, but that we do not need to do this when we dive in headfirst. This is true, but it seems much more likely that, if we did go through a prolonged aquatic phase, we would have taken the more obvious step of evolving nostril valves like a seal. It would only require a small evolutionary step to develop a nose that could shut tight under water. If we did this there would be no need to develop a long nosetip with downturned nostrils – and valvular nostrils would be so much more useful to an aquatic ape.

But perhaps the shape of the human nose helped it to act as a shield of a slightly different kind – a shield against dust and wind-driven dirt. Leaving the peace of the trees and moving out on to the open plains and other more hostile environments, our remote ancestors must have encountered harsh, windy conditions where a down-

■ There are considerable racial differences in nose shape. These are related to the types of environment in which the various human races evolved. Those peoples living in hot dry climates should have tall, projecting noses. Those inhabiting hot damp places should have wider, flatter noses. This would make sense if we consider the nose as a well adapted air-conditioning unit. And this is generally what we find – desert Arabs, for instance, having tall prominent noses and rain-forest dwellers having broad snub noses. There are one or two apparent exceptions to this rule: the Bushmen of the Kalahari Desert and the desert-living tribes of Australian Aborigines both have noses that are too flat and wide. There is a simple explanation for this, however. In both cases, they have only recently been driven into these arid regions, which are not their ancestral homes.

A baby's nose is sometimes so appealing that adults cannot resist touching it.

■ *The male nose is in general larger than that of the female. In this respect the female face is closer to the infantile condition. We know that a flattened snub-nosed face has a powerful parental appeal, arousing strong protective feelings in adults, and during evolution it appears that adult females have exploited this fact by retaining a more childlike profile and thereby arousing the protective feelings of their male partners. The attraction of a tiny nose on a baby is sometimes so irresistible that adults cannot refrain from reaching out to touch it. Experiments with stylized face patterns suggest that this strong positive reaction to a small flat human nose is global and probably inborn in our species. Cartoonists such as Walt Disney have capitalized on this in attempting to design animal and human characters with maximum audience appeal.*

turned nose would have served them well. It is this fourth theory which seems to carry most weight. This looks upon the nose as an air-conditioning plant faced with an increasing burden as our ancestors spread out into the colder and drier regions of the earth. To understand this it is necessary to take a look inside the nose.

When the outside air is inhaled through the nostrils it is hardly ever in an ideal state for passing on to the lungs. The lungs are fussy about the kind of air they get – it must be at a temperature of 95°F (35°C), a humidity of 95 per cent, and it must be free of dust. In other words, it must be warm, wet and clean, to prevent the delicate linings of the lungs from drying out or becoming damaged. The nose achieves this in a remarkable way, supplying over 500 cubic feet of conditioned air regularly every twenty-four hours.

The whole of the inner surface of the complex nasal cavities is covered with a mucous membrane which secretes about two pints of water a day. This damp surface is not static: it is always on the move, because embedded in it are millions of minute hairs called cilia. These keep beating away at a rate of 250 times a minute, shifting the mucous blanket along about half an inch a minute. With the help of gravity, the mucous layer moves down towards the back of the throat, where it is swallowed. It takes about twenty minutes to shunt the mucous layer through the complex system of nasal cavities. While this is happening, the air passing through these cavities gets warmer and warmer and more and more damp. The dust and dirt it is carrying collects on the mucus and is swept away. The lungs are safe for another breath.

By an efficient piece of engineering, the breath exhaled from the lungs loses some of its heat and moisture as it passes back through the nose and into the outside air. About a quarter of the heat and moisture given to the air on its way in is taken again as it leaves. But three-quarters of it is lost, and this becomes dramatically visible as steamy breath when we are out of doors on very cold winter days. On an average day we forfeit about half a pint of water in this way.

If a hospital patient loses the use of his nose for some reason he will find his lungs in serious trouble within only a day or so. Attempts to produce an artificial nose for such a person have run into many

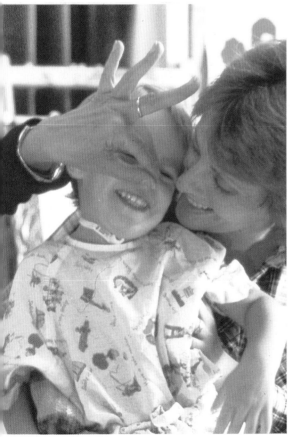

difficulties, underlining the amazing engineering efficiency of the human nose. And it is capable of operating in incredible extremes of climate. Anyone who has experienced a temperature of, say, −30°F, will know that with each inhaled breath the moisture in the lower part of the nose, near the nostrils, freezes solid. Then, with each exhaled breath, it thaws out again. The sensation is far from pleasant, but it is a welcome reminder that the air-conditioning plant in the nose is still able to carry out its task.

A vivid example of the nose's efficiency occurred when a World-War-II fighter pilot lost the hood of his cockpit and was forced to fly

The very best noses hardly exist at all.

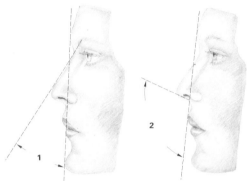

■ *Because we see small noses as infantile and therefore appealing it follows that large protruding noses make females appear ugly. We have no wish to protect the owner of such a nose, since it fails to arouse our parental feelings. A model girl or an actress who possesses an unusually small nose, on the other hand, appears not only appealing but also beautiful. For this reason many females who earn a living by virtue of their otherwise attractive appearance subject themselves to plastic surgery which reduces their naso-facial angle (1) and their columella-lip angle (2) to an appropriate degree. The first of these angles should be between 36 and 40 degrees; the second between 90 and 120 degrees.*

home in freezing air of −25°F at a speed of 250 miles per hour. After two hours of this torture the exposed tissues of his face were destroyed by frostbite, and yet the internal damage was no more than a sore throat. Although it is often ridiculed, the human nose is a truly remarkable organ.

For this air-conditioning function to operate properly, the nose has to be as big as possible. But if it protruded too far it would obscure frontal vision and would be too vulnerable. A compromise has to be reached, and that compromise is the triangular lump of bone, cartilage and flesh in the centre of the face.

There is one way to test this air-conditioning theory. If people live in cold and/or dry regions they should have much larger, more projecting noses than if they inhabit the hot, damp tropics. And this is precisely what we find if we look at the distribution of human noses around the world.

In a *cold dry* climate it has been calculated that only about 1 per cent of the moisture in the breath entering the lungs comes from the outside air; 99 per cent has to be supplied by the nasal linings.

In a *cool moist* climate 9 per cent of the moisture comes from the air and 91 per cent from the nasal linings.

In a *hot dry* climate 27 per cent of the moisture comes from the air, 73 per cent from the nasal linings.

In a *hot moist* climate there is a dramatic difference, with 76 per cent

Males too are restyling their noses.

of moisture now coming from the outside and with the nose being asked to contribute only 24 per cent.

It follows from this that if our ancient ancestors started out in the hot steamy tropics, they would then have had the flat wide noses typical of tropical forest apes such as the gorilla and chimpanzee. If they then moved out on to the hot dry plains in search of game their noses would need some adjustment – an increase in size to step up the power of the air-conditioned unit. With a later spread to the cooler regions of the Earth, further north, the demands on the nose would become even greater and its shape would become even less like the flattened, ape-snout.

Today, of course, no humans have flat, ape-like noses, because all modern human beings, wherever they live, are the descendants of ancestors who went through this climatic shift and developed the larger, typically human nose. But, having spread over much of the Earth's surface, the human race then re-invaded many of its ancient strongholds, so that today we have peoples living in virtually all the different climatic zones. Those that have been inhabiting their present zones for a long period of time do have noses to match. Careful mapping reveals that it is possible to classify people by their nasal index and show how they then fall into regional groups which match up with the temperature and humidity. This does not mean that they are being classified into what are usually called the 'races' of man. For example black people living in hot wet areas in, say, West Africa, will have much wider flatter noses than even blacker people living in the much drier grasslands of East Africa. And white people living in the drier colder regions will have higher, narrower, more beaky noses than white people living in more temperate zones. Nose shape is simply an indication of the kind of air your ancestors breathed and nothing else.

To sum up, then, the human nose is a resonator and a bony shield which grew taller and longer as mankind spread out and away from its hot moist Garden of Eden, keeping its air-conditioning function up to scratch. But there is more to the nose than this, of course, for it is also our main organ of smell and 'taste'. The smelling is carried out by two small odour-detecting patches of cells about the size of a small coin, high up in the nasal passages. These patches are made up of about five million yellowish cells (your dog has 220 million of them) which give us a much better sensitivity to fragrances and odours than we usually realize. Because we cannot sniff the air as well as a dog we seem to think that we are poorly equipped in this respect, but we are not. We are perfectly capable of detecting certain substances in dilutions of less than one part in several billion parts of air. And experiments have proved that the human nose is good enough to be able to follow a fresh trail of invisible human footprints across a 'carpet' of clean blotting paper. The reason we play down our nose-power is because we have increasingly ignored and interfered with its operations. We live in towns and cities where natural fragrances are smothered, we wear clothes that cloy and sour our natural healthy body odours, and we spray our world full of scent-killers and scent-maskers. We even think of 'smelling' as somehow primitive and brutish – an ancient ability best forgotten and left behind. Only in certain specialized areas – the wine-taster, the perfumer – is there

■ *Recently some male performers have also had their noses reduced to alter their appearance. Although this makes them look more infantile and less masculine, it gives them added appeal of a kind that helps to improve their public image. Usually, males only indulge in this kind of cosmetic surgery when they have an exceptionally large nose; but sometimes, as in this case, a nose which is perfectly acceptable as a male appendage is reduced to an almost feminine level, to create a strikingly unusual facial form.*

The phallic symbolism of the larger male nose is based on a popular myth.

any attempt to educate the modern nose and develop its full and extraordinary potential.

I called the nose our main organ of taste as well as smell, and this requires explanation. The tongue is the true organ of taste, but it is very crude in its ability. It can distinguish only four qualities – sweet, sour, bitter and salt. All the other 'tastes' of our widely varied cuisine are in fact detected not on the eager surfaces of our slobbering tongues, as we munch and chew and gulp our meals, but on the small, odour-sensitive patches high up in our nasal cavities. Odour-bearing particles make their way there either directly through the nose as we bring the food to our mouths, or indirectly from the mouth itself. A meal may taste good (on the tongue) but it smells delicious (in the nose).

Because of its association with bad odours, its link with primeval animal 'snouts', and its tendency to dribble and run when we have a cold, the human nose has somehow become the 'joke' organ of the face. We speak of smouldering eyes, delicate cheeks and sensuous lips with awe. But when we go out of our way to refer to the nose it is usually in some derogatory fashion. We have names such as schnozzle, conk, hooter, bugle and snoot with which to insult our nasal miracle of engineering and chemical detection. To be beautiful, a human nose must not possess any particular character; it must be totally *without* any character. A brief glance at the faces of model girls in glossy magazines reveals that the very best noses hardly exist at all. They survive in the exaggerated photographs merely as two tiny nostrils. This state of affairs has become more marked in the present century and it is worth asking why this should be.

To understand the decline of the nose, in terms of beauty assessment, we have to look back at the kind of proboscis with which we pushed our way into the world. As babies we all possess tiny, button noses. As we progress through childhood these small projections grow in proportion to the rest of the face and reach their greatest level with adulthood. So it follows that a small nose = a young nose. Add to this situation a 'cult of youth' and the consequence is clear: the smaller your nose the younger you look.

For the female face the situation is more acute, because men on average have larger noses than women. So, to be youthfully feminine it is doubly important to have a small nose. This is the point at which cosmetic surgery enters the scene. A few women with strangely shaped or outsized noses wear them with pride, but they are rarities. It takes someone with the personality strength of Barbra Streisand to get away with it. Others either suffer in silence or seek surgical aid. The 'nose bob' operation has become increasingly popular in recent decades.

For males, a strong prominent nose is less of a problem since it enhances their masculinity. In earlier days it was almost essential to a man of importance. Edgar Allen Poe went so far as to say that 'A gentleman with a pug nose is a contradiction in terms'. Napoleon Bonaparte declared: 'Give me a man with a good allowance of nose... When I want any good headwork done, I always choose a man, if suitable otherwise, with a long nose.' And Edmund Rostand's famous creation Cyrano de Bergerac would certainly agree: 'My nose is huge! Vile snub-nosed ass, flat-head, let me inform you that I am

Yet parallel changes do occur in both nose and penis during male sexual arousal.

■ *The larger human nose is essentially masculine and is often referred to as a phallic symbol, the popular myth being that the bigger a man's nose the bigger his penis. For this reason an individual who has an unusually long nose – such as a Cyrano de Bergerac – becomes a figure of fun and is subjected to endless ribald remarks. Although in reality there is no connexion between the size of the nose and the penis, these two organs do share one property: during sexual arousal they both become engorged with blood. This makes both of them more swollen and more sensitive. They also become hotter. One industrious researcher went to the length of measuring the nasal temperature of a man who was engaged in love-making and discovered a consistent rise of between 3·5 and 6·5°F caused by the congested blood vessels in the spongy tissue of the nose.*

▼ *Modern heroes tend to have smaller, flatter noses, perhaps because they offer today's assertive females a less threateningly phallic image. Certainly, those males who in the past were considered handsome are by modern standards often exceptionally big-nosed.*

proud of such an appendage, since a big nose is the proper sign of a friendly, good, courteous, witty, liberal, and brave man, such as I am.' Many famous men, from Charles de Gaulle to Schnozzle Durante, would surely have agreed with this but, even so, in modern times the nose of the male hero is shrinking slightly. It is almost as if the strong powerful nose is a sexist threat, an aggressive warrior-shield of a nose and an affront to sexual equality. Small-nosed heart-throbs with boyish grins are the new order of the day.

There may be a deep psychological reason for this. Very large noses are not merely masculine, they are also phallic. The human male only has two long fleshy protuberances on the centre-line of the front of his body. One is his nose and the other his penis. Symbolic equations between the two, either conscious and humorous, or unconscious and serious, are inevitable. They have occurred for centuries and were common in ancient Rome, where the length of a man's nose was said to indicate his virility. In this way the 'Roman Nose' became a special term of praise. Also, symbolically, the punishment for certain sex offences was the amputation of the nose. So, for a modern, post-feminist hero to have too long a nose would suggest not merely undue masculinity but imminent rape. The big nose had to go.

Among tribal societies the nose often takes on a totally different role, and one which is given considerable importance. The nasal passages are envisaged as the 'pathway of the soul'. Without realizing it, we today still subscribe to this view when we say 'Bless you!' after a companion sneezes. The blessing is bestowed because it was thought that the force of the sneeze might expel part of the soul, which would escape through the orifice of the nose. Later, in the Middle Ages, when violent sneezing often accompanied the first stages of epidemic disease, the blessing grew in meaning and it survives today as a relic saying.

In certain tropical tribal societies the treatment of a sick man included the blocking-up of his nose. This was to prevent his soul departing his body as a result of the illness. An Eskimo custom required mourners at a funeral to plug their nostrils with deerskin, hair or hay in order to prevent their own souls following the departed soul of the corpse. In the Celebes a man suffering from a serious illness would have fish-hooks attached to his nostrils, the idea being to hook his soul if it tried to escape, and thus prevent it from leaving his body. In many cultures the nose of the corpse itself is blocked to stop the soul passing through it and there are numerous other examples unearthed by anthropologists which reveal that there has been an amazingly widespread belief in the nose as the departure route for the soul. All are based on the idea that the soul is somehow connected to breathing – to the breath of life. Ordinary breathing through the nose, in and out, keeps a balance and nothing is lost. But with explosive sneezing and the tortured gasps of the dying, the traffic becomes one-way, and superstitious precautions must be taken.

Another ancient belief about the nose was that its shape could be used to determine the true personality of its owner. It was eventually developed into a pseudo-science called physiognomy and in the early nineteenth century there was a special branch of it dealing exclusively with the nose called 'noseology'. There is only one element of truth in the physiognomic argument, and that is so obvious that it has little merit. If a particular individual possesses an unusually ugly nose (of any shape) or an amazingly handsome nose (of any shape), then the appearance of the face in each case is liable to influence the behaviour of companions. Being ridiculed for an ugly nose or loved for a beautiful one will, inevitably, make an impact on the developing personality of the nose's owner. An ugly, ridiculed child will grow up into a different kind of personality from a handsome, popular one. To this extent, nose shape does become connected with adult character, but this is a very different matter from claiming that every subtle difference in nasal contour can be read with precision as an indicator of personality traits.

One important way in which the nose *can* be read is an indicator of changing emotions. Like the whole of the face, the nose is equipped with muscles of expression and we can show our feelings by the movement and postures of our noses – at least to a limited extent. The nose is far less expressive than either the eye or the mouth, but it does have a number of specific signals to offer the onlooker. There is the Nose Wrinkle of disgust, the Nose Twist of distrust, the Nose Twitch of anxiety, the Nose Constriction of distaste, the Nose Flare of anger and fear, the Nose Snort of irritation or repulsion and the Nose Sniff in response to a detected odour. This is a simplification, but it gives an adequate picture of the range of nose signals we have at our disposal. With the addition of snivelling, snoring and sneezing, this is about all the signalling the human nose can do. Complex moods may produce mixed expressions, but these are the basic elements.

■ Nose-to-nose contact is used playfully between adults and children from time to time; as a formal greeting it is largely confined to Maori ceremonial occasions. This action is usually called 'nose rubbing', but in reality there is only a fleeting touch of nosetips, accompanied by a typically western handshake.

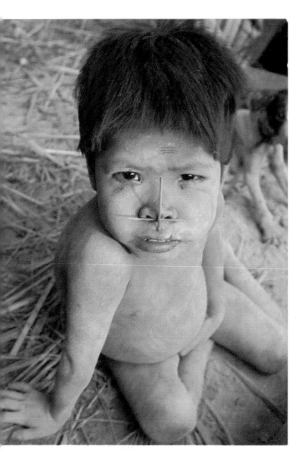

■ Nose ornaments are worn in many different parts of the world. The nose is prepared by having a hole or holes pierced in either the lower part of the septum or the fleshy nostril wings. For young males it has been suggested that these nasal mutilations are examples of 'displaced circumcision'. In all cases, however, one thing is certain: permanent mutilation of the nose is more than mere decoration. It stamps a lasting mark on the owners of the nose, labelling them forever members of their particular group or society.

We also make contact with our noses in a variety of ways. Using our hands, we may touch or rub our noses when we are being deceitful, pinch our nose-bridges when we are deep in thought over some conflicting idea or in a state of exhaustion, or pick our noses when we are bored and frustrated. These are all auto-contact signals which are signs of self-comfort. In one way or another they all signify that the nose-contacting individual is momentarily in need of a little help and opts to provide it for himself with the reassuring touch of his own hand or fingers.

If we are asked a difficult question and wish to hide our inner turmoil while we try to find a suitable answer, the hand shoots up to the nose and touches it, rubs it, grasps it, or presses it. It seems as if the moment of conflict produces a stress reaction in the delicate nasal tissues leading to an almost imperceptible tingling or itching and the hand rushes to the rescue, fondling the nose to calm it down. This is especially noticeable when inexpert liars are telling untruths, and to the trained eye the nose actions are an immediate giveaway.

The pinching of the nose-bridge when deep in thought probably has a similar basis, with the nasal sinuses beneath the bridge causing temporary pain of a mild kind as part of the nose's response to stress. The action of pressing the bridge between the fingers would tend to relieve this pain or at least respond to its presence.

Nose-picking suffers from a mild cultural taboo today, but whenever individuals are alone, or feel that they are so, they may express their boredom or frustration by the trivial comfort action of nose-cleaning. Drivers in traffic jams or waiting at traffic lights are part-

► Some animals have nose-valves that enable them
to close their nostrils at will and, at first sight, it
might appear to be a defect of human nasal design
that we lack this feature. There are so many noxious
smells about today that we would find a closeable
nose a great advantage. These children encounter-
ing tear gas in a Northern Ireland street would
certainly have benefited from this ability. Unfortunately we evolved at a time when dangerous
chemical smells were nonexistent, so that there
was no pressure on us to improve our noses in
this direction.

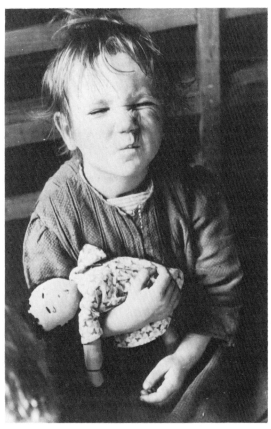

▲ The nose wrinkle of disgust has ceased to be a
response merely to unpleasant smells and has now
become an expressive movement signalling any
kind of repugnance.

icularly prone to frustration nose-picking, providing they are not
carrying passengers.

Far from being gentle and friendly, interpersonal contacts with
the nose in Europe have usually been brutish and nasty. The best a
nose could hope for would be a tweak or a punch. The worst was a
particularly savage form of punishment – the nose slit – in which a
knife was inserted into the nostrils and the nose then sliced open. This
was introduced in the ninth century as a way of dealing with those
who failed to pay their taxes. Although today tax collectors have put
away their knives, we still have a relic of their early methods, in the
popular saying 'to pay through the nose'.

Only between lovers in private was the nose offered more gentle
touches in the western world. During love-making, nuzzling, nose-
pressing and nose-kissing have always been common, but they have
never developed outside the context of sexual intimacy. Among
Pacific islanders they occur in both sexual and non-sexual situations.
Here is Malinowski's translation of a Trobriand man's description of
his love-making: '...I embrace her, I hug her with my whole body, I
rub noses with her. We suck each other's lower lip, so that we are stir-
red by passion. We suck each other's tongues, we bite each other's
noses, we bite each others chins, we bite cheeks and caress the armpit
and the groin...'

In purely social contexts the peoples of the Pacific region employed nose-to-nose contact in much the same way that we would use a
social kiss. Their action is usually referred to as 'nose-rubbing', but
this is an error. The rubbing movements are normally reserved for
the erotic encounters of the kind described by Malinowski. In public
the action is little more than nosetip touching or pressing. It is based
on the concept of mutual smelling with each nose inhaling the fragrance of the other's body.

As a formal greeting, nose-touching is sometimes subjected to
rigid rules of status. In one culture, the Tikopia, there is a whole
range of parts of the body which may or may not be touched with the
greeter's nose. Nose-to-nose or nose-to-cheek contact is only allowed between social equals. When a junior meets a senior the contact
must be nose-to-wrist. When a follower greets a great chief, it must be
nose-to-knee.

Nose greetings are on the decline today. The more cosmopolitan
way of life, increased travel and mixing of cultures, more and more
tourism and international trade, have all contributed to a greater
uniformity of greeting gesture, with the ubiquitous handshake
spreading to cover almost the entire globe. Nowadays, when high-
ranking Maoris meet they combine a vigorous handshake with a
fleeting nose-touch – the new edging out the old.

THE EARS

■ The visible part of the human ear is a rather modest affair. During the course of evolution it has lost its long pointed tip and its mobility. Its fine, sensitive edges have gone, too, curled over into a 'rolled rim'. But it should certainly not be dismissed as a useless remnant.

The main function of the external ear remains that of a sound-gatherer – a flesh-and-blood ear-trumpet. We may not be able to prick our ears like other animals, or twist and turn them when seeking the direction of a sudden noise, but we are still capable of detecting the source of a sound to within 3 degrees. What we have lost in ear mobility we have made up for with head mobility. When a deer or an antelope hears an alarming sound, it raises its head and twists its ears this way and that. When we hear such a sound, we turn our heads, and it works almost as well.

Although our ears feel so rigid at the sides of the head, they do retain a ghost of the movements they once boasted. If you strain the muscles of the ear region tight, while looking in a mirror, you can just see a trace of the protective movement, as your ear tries to flatten itself against the side of your head. Animals with big mobile ears nearly always flatten them when fighting – trying to keep them out of harm's way – and we still do this automatically when we tighten the head skin in moments of panic, even though our ears are already flattened in their normal resting posture.

The shape of our external ears is important in delivering undistorted sound to our eardrums. If we were unfortunate enough to have our ears sliced off by kidnappers to encourage our relatives to disgorge funds for our release, we would find that after our rescue we would be far less efficient as listeners. Our ear canals and their eardrums would form a 'resonant system' and we would find some sounds being emphasized at the expense of others. The seemingly haphazard shape of the ear – its curving folds and ridges – is in reality a special design preventing any distortion of this kind.

A minor function of our ears is temperature control. Elephants flap their huge ears when they are overheated and this helps to cool the animals down. There is a profusion of blood vessels near the surface of the skin and heat loss by this route can be important to many species. For us it may only play a trivial role in thermo-regulation, but it has become a social signal. When someone overheats in a moment of emotional conflict, their ears may go bright red. This ear-blushing has been the subject of comment since ancient times. Nearly two thousand years ago, Pliny wrote: 'When our ears glow and tingle, someone is talking of us in our absence.' And Shakespeare makes Beatrice ask the question 'What fire is in mine ears?' when she is being discussed by others.

Finally, our ears appear to have acquired a new erotic function with the development of soft fleshy lobes. These are absent in our nearest relatives and appear to be a uniquely human feature, evolved as part of our increased sexuality. Early anatomists dismissed them as functionless: 'a new feature which apparently serves no useful purpose, unless it is pierced for the carrying of ornaments'; but recent observations of sexual behaviour have revealed that during intense arousal the earlobes become swollen and engorged with blood. This makes them unusually sensitive to touch. Caressing, sucking and kissing of the lobes during love-making acts as a strong sexual stimu-

▼ *Human ears are greatly reduced in mobility, compared with those of other primates, but they still display vestiges of their former glory. Anatomists have found that there are traces of nine muscles around our ears to remind us of earlier, ear-twitching times. Even today some individuals, after much practice, are capable of quite impressive ear-waggling. Another vestigial element noticeable in about one in four people is a small bump on the rim of the ear, called 'Darwin's Point'. Charles Darwin considered this to be a last surviving indication that our ears were once long and pointed.*

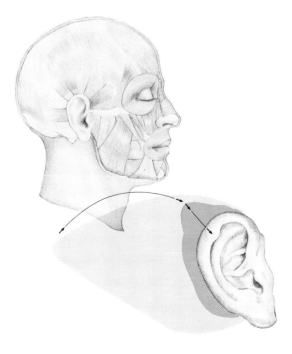

■ No two people have exactly the same ear pattern and it was once thought that all criminals should be labelled according to their ear-type. A major variation concerns the attachment of the ear-lobes. Out of every three Europeans, two will have unattached, hanging lobes and one with have smoothly attached lobes. There are also many smaller variations in the arrangement of the ear folds, each of us possessing a personal 'ear map'.

lus for many people. In rare instances, according to Kinsey and his colleagues at the Institute for Sex Research in Indiana, 'a female or male may reach orgasm as a result of stimulation of the ears.'

At the centre of the external ear is the shadowy 'earhole' which leads to a narrow canal about an inch long. The canal twists slightly, giving it a design that helps to keep the air inside it warm. This warmth is important for the proper functioning of the eardrum at its inner end. The eardrum itself is an extremely delicate organ, and the canal not only keeps it snugly warm but also protects it from physical damage. The price we must pay for this protection, however, is the presence on our bodies of a deep recess which we cannot clean with our fingers. We can groom the rest of our bodies with comparative ease, freeing ourselves from dirt and small parasites, and we can snort, sniff or blow our noses to cleanse those other open recesses, the nostrils; but if an invader enters one of our ear canals, we are in trouble. Attempts to remove the irritation with a thin stick can easily damage the eardrum and we clearly need some special defence against intrusions of this kind. Evolution has provided the answer in the shape of hairs to keep out larger insects, and ear-wax to defeat smaller creatures. The orange-coloured ear-wax has a bitter taste which is repellent to insects. This wax is produced by 4,000 tiny ceruminous glands which are highly modified apocrine glands – the kind which produce the strong smelling sweat in the regions of the armpits and the crotch. As often happens with such evolutionary strategies, the parasites have fought back and certain mites have developed a resistance to the repulsive nature of the sticky foul-tasting wax.

One curious aspect of this wax is that it differs in consistency between one racial group and another. All negroid peoples have the

▲ Human ears are thicker and fleshier than those of other primates. During love-play they become engorged with blood and as sexual arousal proceeds they are increasingly sensitive to touch. If kissed, stroked with the tongue or gently nibbled they act as additional erotic zones. In this respect they reflect an evolutionary shift in the preferred sexual posture of the human species, which makes facial contacts more frequent.

▼ *Early artists always portrayed the Devil with pointed ears. Today the Devil has become a joke figure, little more than a bogey-man to frighten small children, but in the realms of science fiction pricked ears linger on as the symbol of an alien being.*

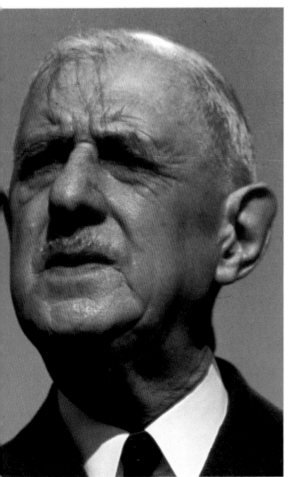

sticky version, as do most caucasoids, but some of the latter and all mongoloid peoples produce a dry wax. Both kinds of wax taste bad, but the sticky form also provides an unpleasant surface which aids the defence of the ear canal against unwanted settlers. It is not at all clear why black and white races should need more protection than orientals.

This is not the place to tell the detailed story of the interior parts of the ear. Briefly, the sound vibrations which strike the eardrum are converted into nervous impulses for transmission to the brain. The eardrum is incredibly sensitive, capable of detecting a vibration so faint that it only displaces the surface of the drum a thousand-millionth of a centimetre. This displacement is then transmitted through

three strangely-shaped bones (the hammer, the anvil and the stirrup) in the middle ear, which amplify the pressure 22 times. The enlarged signal is then passed on to the inner ear where a strange, snail-shaped organ filled with fluid is activated. Vibrations are set up in this fluid which impinge on hair-like nerve cells. There are thousands of these nerve cells – each one tuned to a particular vibration – and they send their messages to the brain via the auditory nerve.

The inner ear also contains vital organs of balance, three semi-circular canals, one of which deals with up-and-down movements, one with forward movements and one with side-to-side movements. The importance of these organs grew dramatically when our ancestors first stood up on their hind legs and adopted bipedal locomotion. An animal standing on four legs is reasonably stable, but vertical living creates an almost non-stop demand for subtle balancing adjustments. Although we take these balancing organs for granted, they are in fact more crucial to our survival than the parts of the ear which deal with sounds. A deaf man can survive more easily than a man who has completely lost his sense of balance.

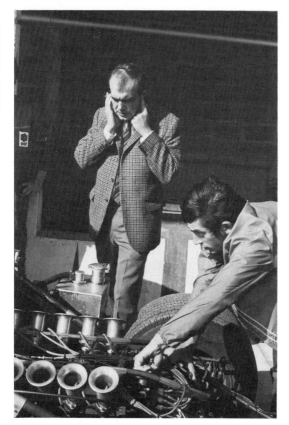

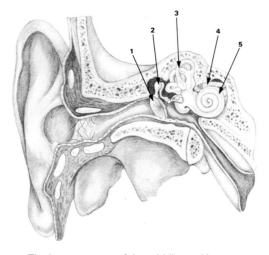

Our ears evolved in a quieter world.

One of the sad aspects of our sense of hearing is that it starts to go into decline as soon as we are born. The human infant can detect sound-wave frequencies from 16 cycles a second up to 30,000. At adolescence, the upper limit has already dropped to 20,000 cycles a second. By the age of 60 this has declined to about 12,000 and the upper pitch that we can detect continues to fall further and further as we become more elderly. For the very old, it becomes a problem to listen to conversation in a crowded room, although they may still be happy enough listening to a single voice in a quiet place. This is because, with their greatly narrowed range of hearing, they find it hard to distinguish between different voices, when several are speaking at once.

Modern hi-fi systems are efficient up to frequencies of 20,000 cycles a second, and it is a galling thought for a middle-aged man who has just spent a large sum of money to install such a system that the only members of his family who will be able to appreciate its full range will be his young children. He himself will be lucky if he can detect its output at anything above 15,000 cycles a second.

Our ears have one serious weakness in relation to volume of sound. We evolved, like other species, in a comparatively quiet world, where the loudest sounds we heard were roars and screams. There was nothing louder to damage our sensitive eardrums and so we developed no special protection against very loud sounds. Today, thanks to our infinite ingenuity, we have thundering machines, high explosives and a whole variety of super-sounds which can very easily damage our hearing.

This weakness of the human ear could easily be exploited by a hostile military power. They would only have to produce a sound with a volume of 185 decibels to burst the eardrums of the enemy forces. A prolonged noise of only 150 decibels would produce permanent deafness. It has also been claimed that exposure to 200 decibels would prove fatal, although this has been disputed. Clearly, our ears serve as a reminder to us all that we now live in a very different world from the one in which we evolved.

Returning to the external ear, it has long been argued that it is possible to identify every individual by his or her ear shape. In the last century it was suggested that this feature could be used to detect criminals, but a rival method – finger-printing – won the day and ear-typing was forgotten. It remains true, however, that it is impossible to find two people with precisely the same ear details. Thirteen ear-zones have been labelled and two of them deserve special mention.

The first is the fleshy lobe. Apart from variations in its size, this has one major classificatory feature. Each of us has either 'free' lobes or 'attached' lobes. Free lobes hang down slightly from their point of contact with the head; attached lobes do not. A doctor who went to the trouble of examining 4,171 European ears discovered that 64 per cent of them had free lobes and 36 per cent attached lobes.

The second is a small bump on the rim of the ear, called Darwin's Point. It is present in most ears, but is often so slight that it is hard to detect. If you feel down the inside edge of the folded rim of your ear, starting at the top, you will come across it about a third of the way down the height of your ear. It feels like a slight swelling, not much more than a swollen pimple, but Darwin was convinced that it was a significant remnant of our primeval days, when we had long pointed

■ *We live in a much noisier world than our ancient ancestors and our ears are not adapted to it. Exposure to too much noise can cause giddiness and nausea, especially if the din is prolonged. Volume of sound is measured in decibels. A whisper is about 20 decibels; ordinary conversation about 60; painful noise in a factory is 100, and discotheques often exceed this level. A gun fired near your ear is in the region of 160. Brief exposure to very loud sound causes temporary deafness. Prolonged exposure can easily lead to permanent deafness. Anyone attending a particularly noisy disco or rock concert will have slightly impaired hearing throughout the following day, during which time their ability to concentrate will be below par. Such is the sensitivity of the much abused ear.*

▲ *The key structures of the middle- and inner-ear are: (1) the eardrum; (2) the trio of tiny bones, named the hammer, anvil and stirrup, respectively; (3) the three semicircular canals; (4) the auditory nerve; and (5) the snail-shaped cochlea.*

▲ *Modern kidnappers have developed the nasty habit of severing a victim's ear and posting it to his family as a means of tormenting them into paying ransom money. This occurred in the famous Getty kidnapping in Italy in the early 1970s, when the victim's right ear was mutilated. Those unfortunate enough to suffer this fate will have discovered that, despite their small size, the ears are excellent sound-collectors and that their special design of ridges and folds helps to prevent distortion as sounds enter the ear canal.*

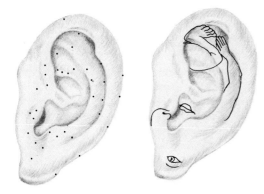

▲ *In the strange symbolism of acupressure and acupuncture, the shape of the ear is seen as concealing an inverted embryo. This means that the embryo's eye, for example, is the centre of the earlobe, from which it follows that, if you are suffering from an ailment of the eye, you must have the finger pressure or the acupuncture needle applied to your earlobe.*

ears that could be moved about freely, searching for small sounds. In his own words, these 'points are vestiges of the tips of formerly erect and pointed ears.' Careful studies have revealed that they are present in a conspicuous form in about 26 per cent of Europeans.

It is variations in details like these that make the ears a suitable subject for criminal identification, but the use of finger-printing has now reached such an advanced state that it is doubtful whether ear shapes will ever be needed. Unhappily the only people to make elaborate studies of ear zones at present are latter-day physiognomists with their romantic claims of character and personality identification from the reading of facial proportions. Their fanciful comments, which lost all credibility earlier this century, have astonishingly resurfaced in the 1980s. As late as 1982 it was possible to read that a large ear is the ear of an achiever; that a small, well-shaped ear belongs to a conformist; and that a pointed ear is that of an opportunist. These and hundreds of other such 'readings', often going into great detail, are an insult to human intelligence and their current popularity is hard to understand.

Criminologists studying facial details have reported that ear design can never be predicted from facial design. If you see a rounded face and an angular face it is impossible to predict which will have the more rounded or angular ear shapes. Somatotyping experts do not quite agree with this. They claim that endomorphs (the more roly-poly among us) and ectomorphs (the more bony and angular) do have different types of ears. Endomorph ears are characterized as lying flat against the head and having the lobe and the pinna (the flap of the ear) equally well developed. Ectomorph ears, by contrast, have the pinnae projecting laterally and better developed than the lobes. Perhaps the explanation for this disagreement is that the criminologists are only considering the head region, while the somatotypers are examining the whole body shape.

Symbolically, the ear has been given several roles. Because it is a flap of skin surrounding an orifice it has inevitably been viewed as a symbol of the female genitals. In Yugoslavia, for instance, a slang expression for the vulva is 'the ears between the legs'. In some cultures mutilation of the ears was employed as a substitute for female circumcision. In parts of the orient young girls at puberty were forced to go through an initiation ritual in which holes were bored in their ears. In ancient Egypt the punishment for an adulteress was the removal of her ears with a sharp knife – another example of the ears as genital substitutes.

Because the ears were seen as female genitals in many different cultures, it is not particularly surprising to discover that certain exceptional individuals were born through the ear. Karna, the son of the Hindu sun god, Surya, emerged from this organ and so, in some legends, did the original Buddha. In the works of Rabelais, Gargantua also enters the world in this unusual fashion.

A completely different form of ear symbolism sees this organ as representing wisdom. This is because it is the ear which hears the word of God. This has been given as the excuse for pulling the ears of children when they are naughty, the idea being that the activation of the ear will wake up the intelligence lying dormant there.

A more curious belief was that the boneless fleshy earlobe had

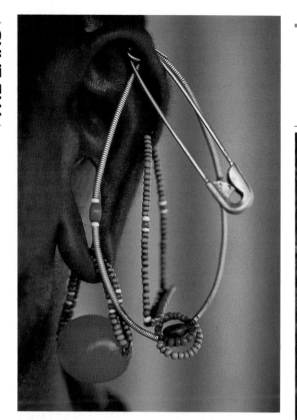

The multi-pierced ear is now found worldwide.

■ *Earrings have been worn since the Early Bronze Age – for over 4,000 years and remain today the most popular and widespread form of voluntary body mutilation. Originally they were worn to protect the wearer from some imagined danger, but their role as 'lucky charms' soon became secondary to their use as high-status-display objects demonstrating wealth and social standing. In this context they often became so cumbersome and heavy that the boneless flesh of the ear suffered considerably. In recent times the multiple piercing of the ears has been adopted as a social badge by young urban rebels reacting against the 'comfort doctrine' of the Armchair Age.*

■ *The reappearance of the male earring in the 1980s has been strongly criticized by the older generation. They consider it effeminate and one school head recently insisted on covering the pierced ears of his male pupils with sticking plaster while they were in class, to hide the offending earrings from view. Such attitudes ignore the long tradition of male earrings. Even Shakespeare (above) wore one. The original reason for wearing a single earring is superstitious. There was an old belief that if a man and his wife or sweetheart each wore one half of a matched pair they would always be reunited, no matter how far they were separated by the man's travels. This made them particularly popular among early sailors who repeatedly embarked on long and hazardous trips. By wearing half a pair of earrings they felt that their chances of returning safely to their loved ones would be increased. Pirates also loved to sport a single gold ring in one ear, but they were less renowned for their fidelity to a single loved one. In their case the explanation was different. They wore a heavy gold band so that if they were killed in some foreign place or swept overboard and drowned, to be washed up on some strange shore, there would be a sufficient weight of gold punched securely into their ear to provide a decent burial. Gold carried in any way could easily be stolen or mislaid, but a thief's attempt to remove a band of gold from a pierced ear would disturb even the heaviest of drunken slumbers. This may explain why it is that in certain parts of the world thieves were punished by having their ears cut off.*

some connexion with the boneless fleshy human penis. Tutors of royal children in the orient, who were not permitted to punish young princes in a more direct physical way, were allowed to pull their ears when they misbehaved, because it was thought that, by so doing, they would not merely be chastising them, but would also be helping to elongate their penises and thus provide them with increased sexual vigour.

Some of these strange superstitions were to lead to the ancient custom of piercing the ears for earrings. This primitive form of mutilation has proved remarkably tenacious and is one of the few types of artificial deformity that has retained a widespread popularity in the modern world. Today most people who have their ears pierced do so for purely decorative purposes, with no understanding of what this once meant. In ancient times there were several explanations:

Because the devil and other evil spirits are always attempting to enter the human body to take it over, it is necessary to protect all the orifices through which they might gain access. The wearing of lucky charms on the ears is thought to have been the best way to guard against an earful of demons.

Because ears are the seat of wisdom, it follows that the very wise have very big ears, especially earlobes. Heavy earrings which pull down the lobes and make them even longer must therefore increase natural wisdom and intelligence. It was reported that Buddha possessed particularly generous lobes, in keeping with his greatness. A study of Hindu, Buddhist and Chinese sculptured figures from earlier days reveals that, if they were important royal persons, they always possessed elongated earlobes.

For no particular reason, sailors believed that to wear earrings in pierced ears would somehow protect them from drowning. They were also convinced that earrings would cure bad eyesight. It is hard to see how such superstitions could arise or persist, but persist they did to the point where sailors and male earrings became almost synonymous.

Over a long period of time these original, varied reasons for donning earrings have been forgotten. The custom becomes referred to simply as traditional. Today nearly all earrings, both tribal and urban, are worn for reasons of status or beauty. In countries where long earlobes are fashionable, the process of mutilation often starts out in infancy, with tiny children having holes bored in their earlobes. These small holes are then gradually enlarged, year by year, so that the ear stretches further and further downwards. By puberty, only the girls with the longest ears are considered to be beautiful. The really gorgeous ones must have ears down to their breasts. If, in the process, the long hanging loop of ear-flesh snaps under the strain of repeated tugging and the weight of heavy ear ornaments, a girl's beauty is instantly ruined. Where this type of exaggerated ear display is the province of the male, a snapped earlobe is less of a disaster, the male in question simply tying the loose ends in a bow.

In the western world, earrings, so long a purely female adornment, have recently been seen on increasing numbers of male ears. At first it was assumed that the wearers were all effeminate homosexuals, but it soon became clear that the habit was spreading to the more avant-garde of the young heterosexuals. This led to some con-

▲ There are few regional gestures involving the ears. A familiar one is the 'donkey's ears' action in which the thumb-tips are pressed into the ears and the vertically spread fingers are waggled derisively at a taunted victim. This is popular as a childish insult in many countries, but is rarely used by adults as a threat, for fear of appearing infantile. A more hostile version does exist, however, in the Middle East. The difference is that when adult Arabs make the gesture they place their little fingers, rather than their thumbs, in their ears. The palms face backwards instead of forwards, and the fingers are spread out vertically, giving a rough impression of a deer with antlers. The message of this gesture in such countries as Syria, Lebanon and Saudi Arabia is that the recipient of the signal is a cuckold.

▲ The ostentatious public use of the confidential word-in-your-ear gesture is a favourite 'excluding' affectation of the arch and the self-important.

The complicated world of ear-touch gestures.

fusion and stories began to circulate that there was a secret code – that to wear an earring in a pierced left ear was homosexual, in a pierced right ear rebel heterosexual. Or vice versa – that was the problem – nobody could ever remember which was supposed to be which. In the end the male earring lost its sexual significance altogether and simply became a generalized way of annoying middle-aged puritans.

During the brief flowering of punk rock in the 1970s, the outrage factor was magnified by the bizarre nature of the objects thrust into the clumsily drilled earlobes. Large safety pins were clear favourites, but chains carrying everything from razor blades to electric light bulbs were also used by the shock troops of the new wave.

In the 1980s the male earring has spread its range even further, despite mutterings from the corridors of high fashion. Even top footballers have been observed signing lucrative new contracts with a fancy diamond stud glinting from one macho earlobe. In adopting this fashion, they appear to be reasserting the young warrior's right to display body ornaments as contrived as any female's. Both males and females can now be seen with the latest development in this trend – multi-piercing. Instead of one hole, the ear is drilled again and again, all around its perimeter, so that a whole series of earrings can be attached to it. This is a device well known from certain tribal cultures, but previously unrecorded in western urban societies.

Turning to gestures and actions involving the ears, it has to be admitted that the repertoire is severely limited. We cover our ears to reduce noise and we cup them to increase sound. We rub them or tug at them when we are indecisive and cannot think what to say; and when we are alone we often probe them with the little finger (appropriately called the ear-finger in some countries) in abortive attempts to clean them.

The most interesting ear gesture is the simple earlobe touch. This has many meanings in different countries. Sometimes the gesturer holds the lobe lightly between thumb and forefinger, sometimes he tugs at it and sometimes he flicks it with his forefinger. In certain countries, such as Italy and Yugoslavia, to make any of these actions towards a man is extremely dangerous because it implies that he is effeminate (and should be wearing earrings). In Portugal the message is very different. There it indicates that something is particularly good or delicious and may be used to describe everything from girls to food. Clearly, Italians in Portugal might find themselves confused by the reactions to their insult ear-touch, while Portuguese in Italy might be surprised to find themselves in hospital after employing their ear-touch of praise. A Spaniard would have another interpretation altogether. To him the ear-touch would mean that someone was a sponger – a nuisance who hovers around in bars cadging drinks but never paying for his own. He leaves his friends 'hanging like an ear-lobe'. In Greece and Turkey, the ear-touch usually means that the performer of the gesture will pull your own ear if you are not careful. It is a warning to children of punishment to come. In Malta it signals that someone is an informer – that he is 'all ears', so watch what you are saying. In Scotland, it is a gesture that registers disbelief – 'I can't believe my ears.' This is a multi-message gesture, employing a variety of forms of ear symbolism, each association growing up in certain regions and remaining virtually unknown elsewhere.

THE CHEEKS

■ Since ancient times the soft, smooth, naked cheeks have been thought of as the focal point of human beauty, innocence and modesty. This is partly because the exaggerated roundness of the cheeks of a baby – a feature unique to humans – acts as a powerful infantile stimulus releasing strong feelings of parental love. This early connexion between smooth cheeks and intense love leaves a residue in our adult relationships. In our more tender moments we reach out to touch, kiss, stroke or gently pinch the cheeks of a loved one, homing in on this part of the anatomy because of its associations with the pure love between parent and child. Just as the young mother presses her infant's cheek gently against her own, so do lovers dance cheek to cheek and old friends kiss and embrace cheek to cheek. Symbolically the cheek is the gentlest part of the whole human body.

The cheek is also the region most likely to expose the true emotions of its owners. For it is here that emotional changes of colour are most conspicuously displayed. The blush of shame or sexual embarrassment begins at the very centres of the cheeks – at two small points which turn a deep red – to be quickly followed by the rest of the surface of the cheek skin and then, if the blush intensifies still further, by other areas of skin such as the neck, the nose, the earlobes and the upper chest. Mark Twain once exclaimed that 'Man is the only animal that blushes. Or needs to' – as if it were the terrible misdeeds of human beings that caused their cheeks to flame red with shame. But this is not the context in which the blush is observed. The typical blusher is young, self-conscious, socially rather shy and usually has nothing much to be ashamed about except personal inexperience and unwanted innocence in an atmosphere of sophisticated knowingness.

The fact that blushing crops up repeatedly in erotic situations makes it look rather like a special sexual display of virginal innocence. The 'blushing bride' is a popular cliché of marriage ceremonies, the blush here being the result of a self-consciousness about the thought that everyone present is privately contemplating the girl's imminent loss of virginity. Because blushing is (or was, before modern sexual education led to a greater openness and frankness on the subject) closely linked to courtship contexts and flirtation moments of very young adults, it has come to be linked with sex appeal. The girl who does not blush is either unaware of her own sexuality or is brazen about it. The girl who blushes when a sexual remark is made is obviously aware of her own sexuality but is still unsophisticated. Therefore it could be argued that blushing is basically a human colour signal denoting virginity. In this connexion it is significant that the girls being offered at ancient slave markets for use in harems fetched much higher prices if they blushed when being paraded before potential buyers.

The cheeks also act as indicators of anger – when they flush bright red. This is a different pattern of reddening, a general diffusing of the colour rather than a spreading from the centre of the cheek. If an angry man is bald it can be seen that the spread of redness extends right over the top of the skull. The mood of the angry man or woman is one of inhibited attack. He or she may issue dire threats, but the red skin indicates that the mood is a frustrated one. The cheeks of the truly aggressive individual turn very pale, almost white, as the blood

▼ The sight of a chubby-cheeked baby acts as a strong stimulus to a parentally-minded adult. The roundness of the cheek region of the infant's face is uniquely human. It is not present in infant monkeys or apes and appears to have been evolved exclusively as a 'releaser' of parental attention.

Rounded, dimpled, smooth and creased cheeks.

is drained away from the skin, ready for immediate action. This is the face of the man who is really likely to leap into the attack at any moment. Similarly, if he or she is intensely frightened the emotion of fear will also make the cheeks blanch, ready for the action of fleeing or of striking out if cornered. It is instructive to study the faces of young policemen struggling to restrain a violent crowd. As you scan the row of helmeted faces, you can pick out the white, the red and the pink, all mixed up in a single line and all experiencing the same threats from the crowd. The white ones are ready to strike out, the red ones

■ *If the human body overheats one of the first signs is a flushing of the cheeks. This occurs when the external temperature rises above normal and also when the internal temperature is suddenly increased, either through fever or emotional excitement. The reddening is caused by the dilation of the blood vessels in the skin, which attempt to reduce and balance body heat by increasing the amount of hot blood near the surface. Among light-skinned races these changes in cheek colour act as important signals of emotional condition. They are beyond conscious control: if we are embarrassed, angry or uncomfortably hot in a social context and wish to conceal the fact our cheeks refuse to lie for us. Our only protection in such circumstances is to cover them in some way. This can be done with a deft hand movement, with heavy cosmetics, or with a deep suntan. Another important feature – particularly of beautiful cheeks, – is their smoothness. The only exception to this are cheek dimples which recall the much dimpled baby's body and were once even supposed to indicate the touch of God's finger. Heavily-creased cheeks, though often expressive of character, are considered un-beautiful because they reflect the loss of skin elasticity that comes with advancing age.*

are angry and flustered and the pink ones have seen it all before. In this way the human cheeks transmit their ancient signals of changing emotional states.

In modern times, the bronzed cheeks of the sunbather offer the status signal of someone who has been able to take time off to lie in the sun on some holiday beach. This is a comparatively recent development. In earlier centuries no high-status young female – or 'young lady of fashion', as she would have been called then – would have been seen dead with suntanned skin. In those days brown skin meant only one thing – peasant toil in the fields. Upper-class young ladies would have looked upon a brown skin as utterly repugnant and would even have taken special measures to avoid the slightest tinge from a stroll in the park, by wearing a sun-shading hat or bonnet, or carrying a parasol.

At certain periods in history this anti-sun attitude led to the whitening of cheeks with the aid of make-up. At other times, when it was thought that rosy cheeks as distinct from bronzed ones were a sign of good health and natural vigour, the centres of the cheeks were painted with rouge. If rouge was not worn, young ladies could be found pinching their cheeks outside an important social gathering to bring the blood to them. 'Blushers' are still popular in female cosmetics to this day, although they tend to come and go year by year as

A beautiful face may sometimes be enhanced by a blemish – renamed a 'beauty spot'.

◄ *The ideal female face shape for the ancient Greeks was that of an egg with its blunt end uppermost. This gave the cheeks a narrowing slope to them and emphasized the smaller, weaker jaw of the female. Today's preferences have turned this classical shape upsidedown, as in this photographer's 'egg-head' fantasy. The reason for this change is a shift towards a more baby-faced female image as the modern ideal. The inverted egg shape creates chubbier cheeks and turns the purring, adult feline into a playful sex kitten.*

▼ *A blemish on the cheek is difficult to disguise; but ladies suffering from moles, warts or dimples in this region were quick to find a solution. It was declared that Venus herself had been born with a natural 'beauty spot' on her cheek and that any lady of fashion who chose to emulate her could therefore only gain in beauty. A small blemish was therefore covered with a black patch or turned into a decorative black spot with a make-up pencil. This form of cheek decoration became so popular that women with perfectly smooth skins joined in and adopted face patches and beauty spots as a purely decorative device. The fashion became so important in the early eighteenth century that the position of beauty spots even developed political significance, with (right wing) Whig ladies decorating the right cheek and (then left wing) Tory ladies decorating the left cheek. The beauty marks themselves ceased to be mere spots and were elaborated into stars, crescents, crowns, lozenges and hearts.*

the fashion houses struggle to keep novelty alive for commercial purposes. This form of make-up carries not only the pseudo-healthy signals but also a suspicion of the teenage blush of innocence as well, giving it a double advantage in sexual contexts.

Apart from colour, the shape of the cheeks is also important. A dimpled cheek has always been considered attractive in Europe because the dimples are said to be the mark made by the impression of God's finger. Dimples do not seem to be particularly common today and they have probably always been rather rare, which may account for the unusual amount of folklore and superstition attached to them. There are many old rhymes and sayings about them, such as 'A dimple in your cheek / Many hearts you will seek...' and 'If you have a dimple in your cheek you will never commit murder.'

Among the early Greeks the shape of the cheeks was also import-ant as a standard of beauty, and the Greeks had a special gesture for it: the Cheek Stroke. This consisted of placing the thumb and fore-finger of one hand high on the cheeks, the thumb on one cheekbone and the forefinger on the other. From this starting point the hand is

▲ *When lovers caress each other's bodies the cheek is a primary target for gentle contact. It is stroked, held, pressed and kissed repeatedly during lengthy bouts of intimate embracing. This interest stems from the loving cheek contacts we all enjoyed as babies in our mothers' arms. It also explains why cheek hugging and cheek-to-cheek contacts are so common during moments of emotional greeting or farewell.*

stroked gently down the cheeks towards the chin and off. During this movement the thumb and forefinger are brought gradually closer towards one another, suggesting a tapering shape to the face. It was this egg-shaped face which the Greeks considered as the ideal of beauty. Modern Greeks still interpret the gesture in this way. To the rest of Europe this same action has rather different meanings. To the majority of people it suggests a thin-faced person and signifies illness or emaciation. It also has some puzzling local meanings. In Yugoslavia, for instance, it is a sign of success. At first sight there is no obvious connexion between stroking the cheeks and the achievement of success, but Yugoslav informants explained that the movement is a mime of the removal of cream from the face. The symbolic equation is: 'he who gorges on cream' = 'he who is successful'.

There are a number of other regional cheek signals. Among Arabs a favourite insult is to puff up the cheeks and then deflate them with a jab of the forefinger, implying that the 'puffed-up' victim has been pricked like a balloon. The gesture is used when the victim is thought to be talking nonsense. In many areas the puffing up of the cheeks is used to signal that someone is fat or overeating, but in such cases the cheeks are not deflated with a forefinger.

The word 'cheeky', meaning impudent, stems from the gesture known as 'tongue-in-cheek' in which the gesturer signals disbelief by pressing his tongue hard into one cheek so that the shape of the cheek is distorted. This originates from the idea that the only way the gesturer can prevent himself from saying something critical is to press it

■ *There is considerable variation in emotional body contact between adult males. In some cultures it is almost taboo and is looked upon as homosexual. In others it is completely uninhibited and even today it is possible to witness two adult heterosexual males indulging in totally non-sexual mouth-to-mouth kissing in moments of emotional greeting, parting or congratulation. But in the majority of modern societies there is a compromise between these two extremes and it is there that the cheek-to-cheek embrace surfaces as the favourite form of intense male-to-male contact.*

Non-sexual cheek-to-cheek kissing between males.

hard into his cheek, to stop him uttering the words which are 'on the tip of his tongue'. To show 'cheek' in this way was considered rude, especially if done by children, and in the early Victorian period the terms 'cheek' and 'cheeky' entered the British language.

Another gesture, largely confined to Italy, is the Cheek Screw. The forefinger is pressed into the cheek and twisted round as if screwing something into the flesh. It is known by almost everyone, from Turin in the north to Sicily and Sardinia in the south of the country. It always has the same meaning: 'Good!' In origin it is a compliment to the chef, signifying that the pasta is '*al dente*' or 'on the tooth'. In other words, the food is cooked to just the right consistency, as suggested by the forefinger pointing at the teeth inside the cheek. But as time passed it became used in a wider and wider context to include anything good. When used of a girl, it is roughly equivalent to the English expression 'Very tasty!'

In southern Spain the same Cheek Screw gesture has a completely different meaning. There it is an indication of an indented, dimpled cheek and is used as a gross insult, signifying that the gesturer considers his companion to be effeminate. Because of this an Italian on holiday in southern Spain complimenting the chef on his food might not get the response he expected.

In Germany this same gesture has yet another meaning: 'Crazy!' It is used by drivers when insulting one another and is a modified form of the well-known and widely-used Temple Screw. It was introduced for a special reason. The police had successfully prosecuted drivers for 'insulting behaviour' when they employed the Temple Screw, which was unmistakable. Their answer was to drop the forefinger down to the cheek level and perform the same 'screw loose' action there, claiming afterwards that they had toothache at the time.

Placing the palms together and then resting one cheek on the back of one hand is a widespread sign meaning 'I am sleepy', based on the fact that the moment which most typifies the activity of sleeping is when one's cheek hits the pillow. It is interesting that when people are tired or bored but have to remain sitting at a desk or table they are most likely to adopt a resting posture in which one hand supports a cheek as if propping up a heavy head. When a lecturer or teacher sees this posture he should realize that he is in trouble. A more obvious sign of boredom is the Cheek Crease, in which one mouth-corner pulls back hard to bunch up the flesh of the cheek. This also signifies disbelief and is essentially a gesture of heavy sarcasm.

In some parts of the Mediterranean pinching your own cheek is a signal that something is excellent or delicious. Almost everywhere this same action, but performed on someone else's cheek, is a sign of affection. It has been used in this way for over two thousand years, having been popular in ancient Rome. It is normally used by adults towards children (who frequently hate it), but may also be performed in a joking way between adults. The Cheek Pat with the palm of the hand is employed as a slightly less irritating alternative, but this too can become annoying when it is performed with too much vigour. In cases of false affection this patting action may easily become magnified into a near-slap, leaving the victim in an awkward state of knowing that he has been insulted but unable to do anything about it because the action is so close to the friendly gesture.

The painting, scarring and skewering of the cheeks.

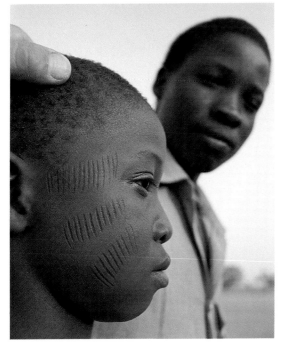

The Cheek Slap itself has a twin tradition. It was the classic way of challenging a man to a duel, and it was also the typical action of a lady responding to the unwelcome attentions of a male. In both cases the slap came as a response to an insult, but the consequences were very different. In essence the Cheek Slap is a 'display-blow' – a blow that makes a great deal of noise but causes so little physical damage that it does not provoke an immediate defensive or aggressive action on the part of the victim. Although it instantly pulls the recipient up short, its significance sinks in later.

At the other end of the emotional scale are the Cheek Kiss, the Cheek Touch and the gentle Cheek Stroke. The Cheek Kiss is a reciprocal action suitable only for two people of equal status. It is a diverted, low-powered Mouth Kiss which has become widespread in western countries as part of the ritual greetings and farewells of social gatherings. Where lipstick is worn it is often more of a cheek-to-cheek press combined with a kissing noise but without lip to cheek contact. There are considerable subcultural variations in its frequency. In theatrical circles and in the more flamboyant social spheres it is almost over-used, whereas in 'lower income' areas it is often extremely rare except between close family. This difference varies as one moves from country to country. In parts of eastern Europe the original mouth-to-mouth greeting kiss, even between adult males, remains common and is not diverted on to the cheek. To some western eyes such actions appear to be homosexual, which is a complete misinterpretation of their significance.

Mutilations of the cheek region have not been particularly popu-

■ *Although popular for cosmetic attention, the cheek region is less frequently used for purposes of permanent scarification and ritual mutilation. This is partly because it is the area of the body most strongly connected with skin smoothness and partly because of its mobility during the many changing facial expressions. Even so, in some tribal cultures cheek scarification can be seen, and the Qadiri Dervishes of the Middle East, for example, still press skewers through their cheeks, apparently without feeling pain, when they have achieved states of ecstasy during their religious ceremonies.*

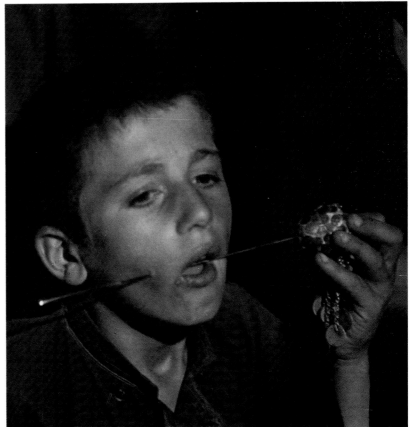

What the cheek pinch signifies depends on the context and the pressure applied.

▶ Gently pinching the cheek of a friend has been considered a sign of affection or a token of congratulation since Classical times; but if performed with too much pressure the gesture quickly becomes an unpleasant assault rather than a loving caress. Many forms of physical contact change their meaning in this way if their strength is increased and there is often a very subtle threshold between the gentle and loving and the powerful and hostile. Sometimes the gentle contact is employed as a deliberate device to lead to an increasingly painful and hostile version.

▲ When a listener rests his cheek on his hand it reveals that he is making an effort to concentrate. This may mean that he is confused, tired or, more probably, bored. Sometimes a listener absent-mindedly pinches his cheek as if trying to keep himself awake. A lecturer who looks up to observe a sea of 'cheek supports' among his audience can be certain that he needs some lessons in the art of public speaking.

lar because of the need for facial mobility in so many situations. In ancient times there was, however, a custom among female mourners of scratching their cheeks and making them bleed as the most obvious way of displaying their agony. John Bulwer reports that this led to a 'keep-cheeks-smooth' law being passed: 'The Roman dames of old were wont to teare and scratch their cheeks in griefe... insomuch, as the Senate taking notice thereof, made an edict against it, commanding that no women should in time to come, rent or scratch their Cheeks, in griefe or sorrow, because the Cheeks are the seat of modesty and shame.'

Prussian students also used to mutilate their cheeks, or have them mutilated, to be able to display proudly their duelling scars. Some scars were actually won in real duels, while others were obtained surgically to give the same impression. In Italy, where razor gangs fought one another in the streets, a slashed cheek also became a symbol of toughness and the membership of an 'élite'. Because of this, there survives to this day a Neapolitan gesture in which a man makes an imaginary cut down his cheek with one of his thumb-nails. He does this to indicate admiration for the cunningness, craftiness or toughness of one of his friends or of a man they are watching. The gesture suggests that the person is a 'scar-carrying member of the gang'.

Tribal decorations of the cheeks include a variety of face-paintings, tattooings, incisings and hole-borings. Apart from the simple powder-and-rouge routines mentioned earlier, the western world is comparatively free of these facial adornments, although there was a brief resurgence of them in the 1970s with the Punk Rock movement in London, when both men and women could be seen with safety pins inserted in the flesh of their cheeks, usually close to the mouth. These savage mutilations of the early punks gradually softened and fake safety pins were eventually put on sale which gave the impression of being skewered through the flesh without actually harming it.

THE MOUTH

The human mouth works overtime. Other animals use their mouths a great deal – to bite, lick, suck, taste, chew, swallow, cough, yawn, snarl, scream and grunt – but we have added to this list. We also use it for talking, whistling, smiling, laughing, kissing and smoking. It is hardly surprising that the mouth has been described as 'the battleground of the face'.

Like any battleground it shows signs of wear and tear. The tongue loses many of its taste buds, obliging the elderly to spice their food; the teeth wear out and need to be replaced; the muscular lips lose their elasticity and slowly settle into a permanently wrinkled condition. The 'relaxed' expressions of the mouths of the elderly give a vital clue as to their personalities. If their lives have been rather glum the constant use of sad mouth expressions will have settled the lips in a permanent downcast mould; if their lives have been full of laughter this too will be reflected in the resting posture of their lips.

The mouth is not only one of the busiest parts of the body; it is also one of the most expressive. This is because changing moods affect the position of the lips in four different ways: open and shut, forward and back, up and down, tense and slack. Combined in different ways these four shifts give us an enormous range of oral expressions. The changes are brought about by a very complex set of muscles which basically operate as follows:

Around the lips is a powerful circular muscle, the *orbicularis oris*, which contracts to close them. It is this muscle that is working hard when we purse our lips or adopt some other tight-lipped expression. It is tempting to think of this as a simple sphincter muscle, but that would be to underestimate it. If the whole muscle contracts, the lips are closed, but if its deeper fibres are activated more strongly its contraction presses the closed lips back against the teeth. If its superficial fibres are more active, then the lips close and protrude forward. So the same muscle, operating in different ways, can produce the softly

As we grow older, the resting position of our lips tends to reflect the emotional mood that has dominated our lives. If we have enjoyed a lifestyle that has been predominantly happy, sad, angry, depressed, enthusiastic or tense, our mouthline becomes set in that expression. An elderly person whose long life has been full of cares and worries may find it difficult to raise the corners of the mouth into a broad grin, even in moments of happiness. The mouth-corners stay stubbornly turned down.

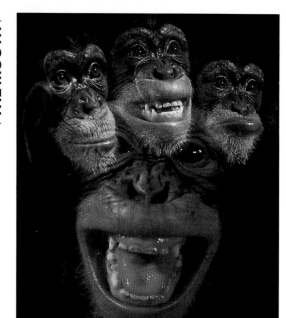

■ *The human mouth is the most expressive in the animal kingdom. Even the mobile lips of our closest relative, the chimpanzee, cannot compete with ours. Studies of infants born blind indicate that we are programmed genetically to make a set of basic mouth expressions related to changing moods; but for the normal child there is considerable refinement of these expressions as part of cultural learning. With combinations of four different contrasts: open and shut, forward and back, up and down, and tense and slack, the lips are capable of developing into a wonderfully subtle visual signalling system.*

The expressiveness of the human mouth is unmatched.

puckered lips of the lover inviting a kiss or the tense, tightened lips of a boxer expecting a punch in the face.

Most of the other oral muscles work against this central circular muscle, struggling to pull the mouth open in one direction or another. To simplify considerably, the *levator* muscles raise the upper lip and help to create expressions of grief and contempt. The *zygomaticus* muscle pulls the mouth up and back in the happy expressions of smiling and laughter. The *triangularis* muscle draws the mouth downwards and backwards in the glum expression of sadness. The *depressor* muscles pull the lower lips down to help form expressions such as disgust and irony. There is also the *levator menti* muscle, which raises the chin and projects the lower lip forward in an expression of defiance, and the *buccinator*, or trumpeter's muscle, which compresses the cheeks against the teeth. We use this not only to blow musical instruments but also to help with the chewing of food. When experiencing acute pain, horror or agonized rage we make use of yet another muscle, the *platysma* of the neck region, which drags the mouth down and sideways as part of the tensing of the neck in anticipation of physical injury.

Complicating matters further are the various vocalizations which accompany our mouth expressions. These add a degree of mouth opening or closing which introduces a new element into the subtleties of facial signalling. Take, for example, the contrasting faces of anger and fear. The key difference is the degree to which the mouth-

corners are drawn back. In anger they push forward as if advancing
on the enemy; in fear they are retracted as if in retreat from attack.
But these opposing movements of the mouth-corners can operate
with the mouth open and noisy or with it closed and silent. In silent
anger the lips are tensely pressed together, with the mouth-corners
forward; in noisy anger – roaring or snarling – the mouth is open exposing both upper and lower front teeth, but again with the mouth-

The extreme mobility of the mouth.

corners in the forward position, making a roughly square-shaped mouth aperture. In silent fear the lips are tensely retracted until they form a wide horizontal slit, with the mouth-corners pulled back as far as they will go; in noisy fear – gasping or screaming – the mouth is opened wide, stretching the lips up and back at the same time. Because fear is retracting the lips, the screamer exposes the teeth far less than the snarler.

Happy faces also have closed and open versions. As the lips pull back and up they may stay in contact with each other, resulting in a wide silent smile. Alternatively they may part to give the broad grin in which the upper teeth are exposed to view. If the sound of laughter is added and the mouth is opened wide the lower teeth may also come into view, but because of the upward curve of the stretched lips these lower teeth are never exposed as fully as the upper ones, no matter how raucous the laughter becomes. If a laughing man *does* expose his lower teeth fully we may doubt the sincerity of his vocal expression.

Another feature of the happy face is the skin-creases which appear between the lips and the cheeks. These diagonal lines, caused by the raising of the mouth corners, are the naso-labial folds, and they vary considerably from individual to individual. They help to 'personalize' our smiles and grins, an important visual factor in strengthening bonds of friendship.

There is one contradictory face, the sad smile, which illustrates another subtlety of human expressions, namely the ability to combine seemingly incompatible elements to transmit complex moods. In the sad smile the whole face composes itself into the twinkly-eyed look of good humour – except the corners of the mouth, which stubbornly refuse to jack themselves up into the appropriately raised position. Instead they droop to create the 'brave smile' of the failed politician or the sardonic smile of the bank manager refusing a loan. There are many other 'mixed' or blended expressions which, together with the single-minded ones, provide the human face with the richest repertoire of visual signals in the animal world.

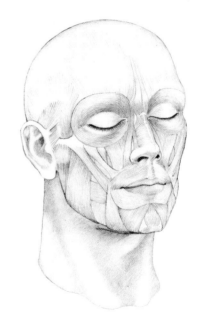

Before leaving the lips to enter the mouth cavity and examine the teeth and the tongue there is one unique feature that demands comment. Human lips are unlike those of all other primate species in that they are strongly everted: they are rolled outwards to expose parts of the mucous membrane. In other species this mucous surface is limited to the inner region of the mouth and is invisible when the mouth is closed. In humans the visible mucous lips are smoother and darker than the facial skin around them, a contrast that helps to make subtle changes in oral expression more conspicuous.

A second function of our exposed mucous lips appears to be sexual. During erotic arousal they become swollen, much redder and more protuberant. This makes them not only more sensitive to contact with the partner's skin but also more conspicuous. The change they undergo mimics closely the alterations that are taking place on those other labia – of the female genitals. Bearing in mind that female lips are slightly bigger than those of the male, it looks as if the evolution of everted lips on the mouth of the human species has been a useful piece of sexual self-mimicry, enabling the female to signal her arousal from the facial region as well as the genital. Since we tend to mate face to face, rather than rump to face as in other primates, this

■ *During evolution the complex facial muscles of human beings have become increasingly adapted to a purely expressive role. This has been possible because a great deal of the heavy work of biting and tearing with the teeth and jaws has been taken over by the hands, employing sharp weapons and tools. This has enabled us to dispense with the massive, fixed mouths found in other species.*

imitative device makes good sense. It also explains why women have for thousands of years painted their lips red to make themselves visually more exciting. It further explains the unconscious factor operating in the attacks on lipstick-wearing by puritanical groups from time to time.

Inside the lips are the teeth which, in our species, are employed almost exclusively for feeding. We may occasionally use them to snip a thread of cotton, but their non-feeding uses are much rarer than in other species. Give an ape a strange object to examine and it will pick it up and almost immediately raise it to its mouth to explore it with lips, tongue and teeth. It may then manipulate it with deft fingers, but overall there is a dependence on both digital and oral contact, with the oral playing the major role. This is also true of human infants, whose parents must always be on the watch for dangerous objects being thrust into tender mouths. As we mature, however, the mouth gradually loses its 'investigative role', which is taken over almost exclusively by our superior hands. This switch also applies to fighting. Apes in a rage grab at their opponents and bite them. Humans in a rage beat their opponents over the head, and punch, kick and wrestle with them. They bite only as a last resort. The same is true of killing prey. Again the hands – with the help of weapons – have taken over the task of the lethal bite so common among carnivores. Along with this shift from mouth to hand, the human teeth have become smaller and are rather modest compared with those of other species. Our canines have ceased to be fangs with long sharp points. They are only slightly longer than the other teeth, with no more than small blunt points to remind us of our distant ancestry.

The full complement of adult human teeth is 32, 28 of which establish themselves by puberty, having gradually replaced the smaller set of 'milk teeth' which we use during childhood. The last four teeth, the wisdom teeth at the back of the mouth, emerge as we become young adults. Sometimes a few of these, or all of them, fail to appear, so that adult mouths can vary in numbers of teeth from 28 to 32.

Apart from their obvious actions of biting and chewing food, teeth are also said to clench and clamp, to gnash, grit and grind, and to chatter with cold. Clenching or clamping the teeth occurs at moments of intense physical effort or when someone is anticipating pain. It is seen on the face of the grappling wrestler or the child about to be injected and is a primeval response to possible injury. If a blow were to fall on the face of an open-jawed individual it could cause much more damage, clashing the teeth together and possibly splintering them or dislocating the 'unclamped' lower jawbone.

To gnash or grit the teeth is the same as to grind them, and it is hard to understand why the language needs three words for the same action, especially as it is so rarely used in real life. During the sleep of many individuals, however, a gentle grinding together of the teeth does apparently take place, indicating a kind of suppressed anger. Again, this is a primitive response which re-surfaces as a sort of 'muscular dreaming', with the frustrated individual symbolically grinding down his enemies in the safety of his slumbers.

Sadly, although tooth enamel is the hardest substance in the entire human body, dental decay is the most common human ailment in the world today. The cause seems obvious enough. A bacterium in the

▲ The strangest feature of human lips is that during the course of evolution they have been turned inside out. Man is the only species of primate with conspicuously everted lips. Because the mucous membrane exposed in this way is darker than the surrounding facial skin, this eversion makes the shape of the mouth more conspicuous and intensifies the expressive oral signals. In dark-skinned races, where there is less contrast between the colour of the lips and the surrounding skin, the exposed mucous membrane has a more clearly defined border – a marked lip-seam – which is an alternative way of making their shape clearly visible.

► During erotic arousal the lips become swollen and redder in colour. Any individual who has lips which are naturally or artificially more protuberent or highly coloured will automatically transmit heightened sexual signals.

mouth which rejoices in the name *Lactobacillus acidophilus* loves carbohydrates, and if particles of sugary or starchy foodstuff are left clinging to the teeth or gums after eating it rapidly ferments them into lactic acid. The bacterium loves this acid even more and starts to reproduce wildly, dramatically increasing the whole process until the saliva in the mouth has become unusually acidic. The acidity then eats away at the surface of the teeth, making small holes in the enamel which develop into festering cavities. All this has been confirmed in a number of ways. For example, children growing up in wartime Europe, when there was very little refined sugar or starch, had fewer cavities. Also, animals fed with a sugar-rich diet develop tooth decay if they eat their food in the usual way but do not if the same diet is tube-fed to them and never touches their teeth. Furthermore, wild chimpanzees living deep in the forest have excellent teeth, while those scavenging near human settlements have rotten teeth.

Yet there are some strange facts about tooth strength which we simply do not understand. Some individuals, for example, seem to be almost immune to decay even when they eat the sickliest of sweet diets. Others fall prey to decay despite great care with both diet and tooth-cleaning. Also, logic would insist that the lower front teeth would, with the help of gravity, become the most food-laden and therefore acid-attacked. Surprisingly, these are the most decay-resistant of all the teeth. In the western world nearly 90 per cent of people have healthy decay-free lower front teeth. In sharp contrast more than 60 per cent have lost their upper-right-middle molar from decay. Despite great advances in dental science the teeth still retain some of their mysteries.

▼ *In western culture only pearly white teeth are considered attractive, but eccentric fashions have occasionally permitted the decoration of a selected tooth with a jewel of some kind, usually a diamond. In some other cultures, however, blackened teeth, or teeth filed to a sharp point, are looked upon as the acme of beauty.*

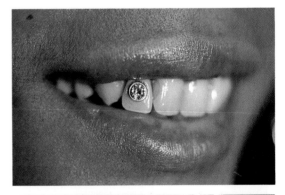

Our canine teeth are no longer lethal fangs.

Lying behind the teeth, the human tongue is a remarkable organ that tastes, masticates, swallows, cleans the mouth, and communicates by gesture and speech. Its rough upper surface is covered in papillae which carry a total of between nine and ten thousand taste buds. These are capable of detecting four tastes: sweet and salt on the tip of the tongue; sour on the sides of the tongue; and bitter at the back of the tongue. It used to be thought that *all* tasting took place on the upper surface of the tongue, but it is now known that this is not the case. There are sweet and salt taste buds elsewhere in the mouth, especially on the upper throat, while the primary tasters of sour and bitter are on the roof of the mouth at the point where the hard palate meets the soft palate.

It is believed that we have these particular taste responses because it was important for our ancestors to be able to tell the ripeness – and therefore sweetness – of fruits; to be able to maintain a correct salt balance; and to be able to avoid certain dangerous foods – which would have tasted very strongly bitter or sour (acidic). All the subtle tastiness of our foods derive from a mixture of these four basic qualities, aided by other flavours which we smell in the nose.

Besides taste, the surface of the tongue is also responsive to food texture, to heat and to pain. During mastication it rolls the food round and round in the mouth, testing it for lumps. When it judges that all sharp pieces have been crushed or rejected it participates in the crucial act of swallowing. To do this its tip presses against the roof of the mouth and then its rear part humps up to catapult the wad of saliva-soaked food into the throat and on its way to the stomach. This extremely complex muscular action is something we take entirely for granted because it is so automatic. It is so basic, in fact, that we were able to perform the action well in advance of any need for it – when we were still inside the womb.

When the meal is over, the tongue busies itself like an outsize toothpick, flicking this way and that, trying to dislodge annoying particles of clinging food from between the teeth. Perhaps it is because the tip of the tongue is nearer to the front lower teeth that they are the cleanest and most decay-resistant.

The tongue is also a visual communication organ – as illustrated on pages 102-3 – and plays a vital part in the complex process of talking. We think of speech as the product of the larynx and tend to overlook the extent of the tongue's involvement. This error is quickly corrected by trying to speak with the tongue held on the floor of the mouth. Anyone who has visited a dentist will have discovered this.

The three main features of the mouth – the lips, the teeth and the tongue – are kept moist by the secretions of three pairs of salivary glands. The pair embedded in the cheeks are known as the parotid glands and they produce about a quarter of the saliva; those beneath the jaw under the molar teeth – the sub-mandibular glands – are the most productive, accounting for about 70 per cent; and those beneath the tongue – the sub-lingual glands – contribute the other five per cent. Estimates of a person's total output of saliva per day vary between one and three pints. More food means more saliva. Fear and intense excitement mean less saliva.

When saliva leaves the ducts of the salivary glands it is free of all bacteria, but by the time it has swirled around the mouth a few times it will have collected up between ten million and a thousand million bacteria per cubic centimetre. It acquires these from the tiny fragments of 'wet dandruff' which are always present inside our mouths, as the skin surfaces there repeatedly slough off old layers and replace them with new tissue.

Saliva has a number of functions. It moistens the food as it enters the mouth and makes it accessible to the taste buds, for dry food cannot be tasted at all. It also lubricates the wad of chewed food before it is swallowed and in this way eases its passage down the oesophagus. Its quality as a lubricant is improved by the presence of a protein called mucin. If food is chewed for any length of time an enzyme in the saliva called ptyalin starts breaking down starch to maltose. Ptyalin also acts as an oral germ-killer, as do other lyzozymes which help to clean the mouth and teeth. Saliva also contains chemicals which create slightly alkaline conditions that help to reduce acid attack on the tooth enamel. Finally, the lubricating action of saliva improves the quality of vocal tones, as anyone who has tried croaking with a dry mouth will appreciate.

One of the most curious actions we perform with our mouths is the yawn. When we are bored or tired we often cannot help stretching our jaws open to the maximum and at the same time taking a deep breath. Anyone observing us is liable to find our yawning action infectious and in no time at all a whole group of people can be set gaping and covering their mouths with their hands. What does it all mean? The truth is that nobody really knows, although there have been some educated guesses. Any suggestion that it has to do with the intake of air is ruled out because fish yawn in water. Could it be a specialized stretching movement involving chest and face muscles? Other stretching actions of the limbs and trunk often accompany it, and the net result is a slight increase in heart-beat which may be part of an attempt by the body to get more blood to the brain. But this seems an unsatisfactory answer. Another possibility is that it is a 'rest synchronizing' action similar to certain pre-roosting actions of birds. Viewed in such a way the yawn becomes more of a visual display – a signal that the yawner is about to go to sleep. Its infectious impact on others would then make good sense. Sadly for this theory solitary animals also yawn, so for the present yawning remains an intriguing mystery.

Less of a mystery is the reason why people place their hands over their mouths when they yawn. This is usually said to be simply a polite way of concealing the inside of the mouth, dating from the time, before modern dentistry, when many adults had rotting black stumps of teeth. This is a plausible explanation but it happens to be wrong. The true explanation goes back much further, to times when it was believed that a man's soul might escape with his breath if the mouth were opened wide enough. The covering of the yawning mouth was intended to prevent the premature departure of the soul – and also prevented evil spirits from taking the opportunity to *enter* the body through the gaping orifice. Some religious sects believed that yawning was a ploy of the Devil and, instead of covering their mouths with their palms, they snapped their fingers as loudly as they could in front of their yawning mouths to frighten away the evil being. Even today in some parts of southern Europe Christians make the sign of the cross when they yawn.

■ *About 20 per cent of adults in the western world have no natural teeth at all in their jaws. For some reason, tooth loss among whites is twice as great as in Negroid peoples. The main problem we face today is that, having extended our average life span by about half a century from the neolithic figure of 22 years, we have demanded a staying power from our teeth which they cannot meet. Like typical mammals we have only two sets of teeth – the milk teeth of childhood (far left) and the optimistically named permanent teeth of adulthood (left). With our longer lives today we badly need a third natural set, but must make do with an artificial one, courtesy of modern dentistry. In this respect human beings are greatly inferior to sharks, where old teeth are replaced by new ones, time after time, throughout life. A full set of human adult teeth includes incisors (1), canines (2), premolars (3) and molars (4). Beneath the enamel outer coating (5) of each tooth is a dentine body (6) and a pulp cavity (7) containing blood-vessels and nerve-endings.*

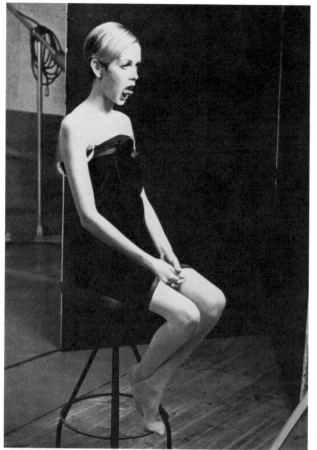

■ *Besides its tasting, feeding and speaking duties the human tongue also transmits several visual messages. These are primarily based on two infantile mouth movements – the stiffly protruded tongue of nipple rejection when the baby has fed enough, and the sinuously exploring tongue when the baby is searching for the nipple. In other words, there is the rejecting tongue and the pleasure-seeking tongue, and these are reflected in the ways in which adults display this normally hidden organ. people who are concentrating hard on some personal task and who do not wish to be disturbed push out their tongues like signs saying 'Busy, keep away'. People who wish to be openly rude also stick out their tongues as an unmistakable rejection gesture. In complete contrast, those who are feeling sexy and wish to signal their desire for a sexual encounter – in which they may literally explore their partners with their tongues – use the sinuous, curling movements of the searching, exploring tongue of infancy.*

In addition, there is a powerful piece of body symbolism which sees the tongue as an echo of the male penis. Obscene mouth gestures often make play with the tongue as a symbolic penis and the lips as symbolic vagina. A common form of invitation from a prostitute, for example, is the slow protrusion of the tongue from between open lips, repeated several times to simulate copulation. And there is a male sexual invitation in South America which involves the slow wagging of the tongue from side to side within the half-open lips.

Mouth-covering gestures when we are not yawning have different origins. For example, during conversation a person may raise the hand to partially cover the mouth, sometimes even keeping it there while speaking. This is a 'cover-up' in both a literal and a symbolic sense and occurs when the person making the gesture is trying to hide something from his or her companions. It is a signal of secrecy, evasiveness or deceit. The hand comes up to the mouth as if to block the words that might issue from the lips. It is wrong, however, to think that such a person must be antagonistic. It may be that all he or she is doing is concealing from you a home truth which might be hurtful to you.

An oral activity which is universally popular is the kiss. Today kissing is used both as a friendly greeting and as a form of sexual stimulation between lovers. In the greeting form the lips are applied to different parts of the other's body according to the relative status of

the kisser and the kissed. If equal-status kissers meet they exchange equal kisses, on the lips or the cheeks. If a low-status kisser meets a high-status being he may kiss the great one's hand, knee or foot, or the hem of his garment. In extreme cases he is allowed to kiss only the dirt near the feet.

The tongue-probing kisses of lovers are much closer to the primeval origins of kissing. It began long ago, as neither a greeting gesture nor a sexual one, but as part of the weaning behaviour of a young mother towards her infant. In the days before there were convenient baby foods the mother had to switch her baby from breast-feeding to solids via morsels which she pre-chewed herself and carefully transferred, mouth-to-mouth, to her growing offspring. This action, which died out only quite recently in the more remote parts of Europe and which can still be observed among certain isolated tribal societies, forged a primitive link between lip-contact and the rewards of being fed by one's mother. Being kissed meant being loved, and offering a kiss meant giving love. This link has remained, even though the weaning process has long since become more sophisticated.

Oral-genital contacts – which we now know are not the modern inventions of 'decadent' Western society but have played a major role in the sexual activities of many cultures for thousands of years – are strongly related to infantile oral pleasures at the breast. When young lovers kiss the clitoris or penis of a partner their mouth movements are strongly reminiscent of those they enjoyed when they were suckled by their mothers. The impression made by the oral stage of life stays with us in some shape or form throughout much of our adulthood.

It should be added that the Freudian view of adult oral pleasures is that they reflect infantile *deprivation*. The suggestion is that infants denied the oral rewards normally provided by mothers will spend the rest of their lives trying to compensate for the loss. In extreme cases this may well be so, but what Freud overlooked is that the pleasures we experience at any stage of life are liable to establish patterns of behaviour for the future. An individual who as a baby enjoyed sucking at the breast, as most do, is hardly likely to forgo the chance of enjoying adult ways of recapturing such a pleasure – simply because there was *no* infantile deprivation. Freud's negative attitude towards adults who enjoy kissing, smoking, eating sweet foods and sipping warm sweet drinks is perhaps not hard to understand because his own mouth caused him endless agony. He suffered from cancer of the palate, most of which had to be removed in a series of 33 operations, so he could be forgiven for his attitude towards adults whom he called orally arrested, breast-fixated and infantile simply because, unlike him, they were able to enjoy adult oral pleasures.

Turning to a completely different kind of oral activity, spitting has a strange history. In ancient times it was considered to be a way of making an offering to the gods. Spittle, because it emerged from the mouth, was thought to contain a small part of the spitter's soul. By offering this precious particle to his supernatural protectors a man could enlist their aid in his endeavours. The danger in this was that if his enemies could collect some of the fallen spittle, they could work hostile magic on it and bewitch the spitter. For this reason some great

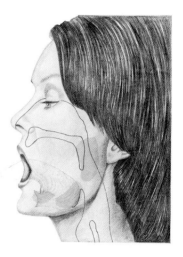

▲ The interior of the mouth is kept moist by the secretions of three pairs of salivary glands (coloured areas).

▲ Yawning is a strange activity during which we stretch our jaw and chest muscles in an exaggerated way. It is often accompanied by the stretching of other parts of the body, with the result that there is a slight rise in heart-beat and perhaps sufficiently increased circulation to improve the blood supply to the brain.

▲ A common form of mouth contact among infants is the thumb-sucking that indicates insecurity. This is an attempt to recapture the comforting feel of oral contact at the breast, when something warm entering the lips meant food and love and security. Infants usually grow out of this habit after a few years, but it resurfaces in the pencil-sucking of students, nail-biting, the sipping of warm sweet drinks when we are not thirsty, the puffing of warm smoke from cigars and cigarettes into our mouths, and the pensive sucking on pipe-stems by elderly males. In all such cases we are reliving the rewarding moments of infancy, when we sucked at our mothers' nipples or the teats of her milk-bottles. The fact that such 'oral comforts' serve to reduce tension in stressed adults is often ignored by medical critics of smoking.

▼ There is now only one sphere (outside the dentist's surgery) in which spitting is tolerated and that is in organized spitting contests – usually involving tobacco, fruit pips or beer. The distances achieved are faithfully recorded in the Guinness Book of Records. The champion long-distance tobacco spit now stands at 45 feet; but with the aid of solid objects this has been far exceeded. Melon pips and cherry stones have both been spat over 65 feet.

tribal leaders employed a full-time spittle burier, whose task was to follow the Great One everywhere with a portable spittoon and bury its contents each day in a secret place.

The magical power of spittle led to its widespread use in the making of oaths and pacts, and spitting on the palms of the hands when making a bargain has persisted in some countries to this day. Fighters who spit on their palms before an encounter are also resorting to this early form of magical protection, although the action has long since been rationalized as 'moistening the palms to get a better grip on an adversary'.

In Mediterranean countries where a belief in the Evil Eye became commonplace spitting was a defence against it. If someone afflicted with the Evil Eye passed by, people would spit on the ground to ward off the dangerous influence. In this way spitting changed from a sacred act into a gross insult. Eventually, spitting *at* someone became a symbolic act of intense hostility, which it remains to this day.

When it comes to projecting things from the mouth, sounds travel farthest. The normal range at which the male voice can be understood is about 200 yards, but the maximum distance at which it has been detected is an amazing ten-and-a-half miles across very still water on a silent night. In certain mountainous regions of the world whistling languages have developed for communication across valleys between people working in the fields. On one of the Canary Islands, La Gomera, there is a whistled language called *silbo*, which is essentially 'whistled Spanish', the pitch and tone variations of the whistles replacing the vibrations of the vocal cords. On a good day these whistled messages can be understood at distances of up to five miles.

Because the mouth is such an important focus of attention when people meet it has inevitably been subjected to many cultural modifications, exaggerations and improvements. Lipstick has already been mentioned, but in various tribal societies lip-tattooing and the insertion of lip plugs have also been employed to alter the facial appearance dramatically.

The insertion of wooden lip-plugs varying in diameter from coin-size to plate-size was widespread among tribal societies in earlier days. These curious adornments were encountered as far apart as tropical Africa and the forests of northern South America. In Africa they were most common among those tribes which had been the main source of female slaves, and the usual explanation given for them was that they made the females of the tribes look unattractive to the Arab slave-raiders. Other stories interpret them simply as marks of high status or as ways of disfiguring females to reduce tribal jealousies. Whatever their true function may have been, they effectively eliminated all forms of oral expression.

In North Africa and parts of the Middle East a less drastic method of restricting facial expressions and eliminating feminine facial beauty in public was the adoption of the veil. In its most extreme form this covered the whole face with only small circular holes left for the eyes. In all its forms it effectively reduced visual communication between married women and strangers.

Western eyes have always regarded a healthy set of gleaming white teeth as an essential mark of beauty, but many cultures have taken a

Weaning behaviour gave rise to kissing.

Timide

Frivole

Passionné

Amoureux

Langoureux

Voluptueux

■ *Kissing originated as a maternal act in which a mother passed masticated food to her infant during the weaning process. Like sucking at the breast, this mouth-to-mouth contact became indelibly linked with infantile comfort and security. It follows that kissing, like sucking, survives into adult life as a comfort act strongly associated with loving relationships. More modest relationships are signalled by modified kisses – on the cheek or the hand. In the past, servile kisses have been demonstrated by making lip-contact while at the same time lowering the body submissively. This led to the development of kissing the hem of a garment, kissing the foot and even kissing the dirt near the foot. Such gross acts of subordination are rare today, one of the few exceptions being the Pope's tarmac-kissing action whenever he arrives in a new country. He employs this form of greeting as a display of personal humility to counterbalance the pomp and power of his official status.*

■ *Huge lip-plugs have been worn in a number of tribal societies, completely disfiguring the faces of the owners and rendering ordinary facial expressions almost impossible. The plugs begin as small, coin-sized discs and are replaced from time to time by slightly larger ones until, with adult-hood, the wearers display massive, plate-sized ornaments from their drooping lower lips.*

different view. One trend has been to remove the central incisors in order to emphasize the pointed canines, which makes the mouth look more menacing and beast-like – almost a Dracula face. This technique has been employed in Africa, Asia and North America.

Another method of making the teeth look savage is to file them to sharp points. This too has occurred over a wide range, from Africa to South-East Asia and the Americas. Sometimes precious stones or metals were inlayed in the teeth to add glamour to them as high-status displays. Many of these tooth operations and mutilations were carried out at special times in the lives of the tribespeople, especially at puberty and at marriage, implying that the mouths were being used symbolically as 'displaced genitals'.

In some areas the impact of teeth was reduced rather than exaggerated. In Bali, for instance, young adults were subjected to painful

▶ *In some Arab cultures, where men guard their wives jealously from the prying eyes of strangers, the mouth is covered by a veil which eliminates any possibility of expressive communication. It also conceals the erotic signals of the female lips and disguises any facial beauty which the veiled woman may possess. Some veils cover only the mouth and jaws, some cover the whole face but leave a slit for the eyes, and some cover even the eyes, leaving only two small, gauze-covered patches for the wearer to see where she is going. If these measures seem extreme, it should be remembered that in some regions around the shores of the Mediter-ranean, until comparatively recently, a female was only allowed out of doors a few times during her entire life – sometimes only twice, to be married and to be buried. For such women the veil was a major advance.*

Although extremely widespread, the mouth shrug is surprisingly not found in all cultures.

tooth-filing to flatten out the points of the canines and make the human mouth look *less* like an animal mouth. In certain other eastern cultures the females blackened their teeth or dyed them dark red, making them virtually disappear from sight and creating an infantile expression, as though they had suddenly regressed to the gummy stage of babyhood. In this way they made themselves appear more subordinate and submissive to their males.

There are many regional gestures involving the mouth. Sometimes the same message is conveyed by slightly different actions as one moves from country to country. For example, the signal for 'Silence!' is usually thought of as the pressing of a raised forefinger to the closed lips, but in Spain and Mexico the message is more likely to be given by holding the lips tightly together between a forefinger and thumb. In parts of South America the signal is made with a thumb-tip moved across from one mouth-corner to the other. In the Bible, silence is requested by the placing of the whole hand over the mouth. In Saudi Arabia the local variant has the forefinger held up near the lips, while the gesturer blows on it.

The signal for food is much the same the world over, with bunched fingers miming the act of pushing food into the mouth. But the signal for drinking has at least two forms. In most countries it simply involves the mime of tipping an imaginary glass up in front of the open mouth; but in Spain a different version dominates, where there is a local custom of drinking from a soft leather bottle which is held up high so that a jet of liquid squirts down into the mouth. To suggest drinking, the Spaniard mimics this action by raising the hand in the air with only the thumb and little finger extended. The other fingers are tightly bent and with them in this position the thumb is jerked downwards towards the open lips. A strange 'descendant' of this gesture, found in Hawaii, is a legacy of the early Spanish sailors who visited the Pacific. Hawaiians employ the same hand posture, with thumb and little finger extended, as a friendly greeting. They no longer direct it to the mouth but instead wave it towards their friends. Most of them have no idea of the origin of this gesture which they make every day without thinking.

Oral gestures indicating anger include the 'Teeth Flick' action of the Mediterranean region, in which the thumb-nail of one hand is placed behind the upper incisors and then violently flicked forward towards the victim. For some reason this gesture is on the decline. It was common as an insult in Northern Europe, including the British Isles, in the seventeenth century, but has since vanished there. Its stronghold today is Greece, but it is also well known in Italy, Spain and southern France. Among some Arabs it is also popular, but others prefer to demonstrate anger by simultaneously biting their lower lips and shaking their heads from side to side, miming the action of a dog killing a rat.

Praise is often indicated by putting the fingertips on the lips and then kissing them towards the object of affection. Smacking the lips together was originally intended as praise only of food but is now often used as a sign of appreciation of a 'tasty' female. Vibrating the lips noisily while exhaling, which once signalled 'hot', is nowadays also used as an appreciative comment on a woman indicating that she is 'hot stuff'.

▼ *There are countless regional mouth gestures with specialized meanings. Some of these operate only in a very restricted area, such as one small island; or they may spread across whole continents. The mouth shrug, performed so eloquently here by a Frenchman, is so widespread that we tend to think of it as a universal gesture for the human species. But this is an error. In the Orient the mouth shrug is almost non-existent as revealed by a recent observation that the more sophisticated, young, westernized Japanese are just beginning to shrug. Even among westerners, the use of the mouth shrug varies considerably from culture to culture, the most exaggerated version being confined largely to Latin countries.*

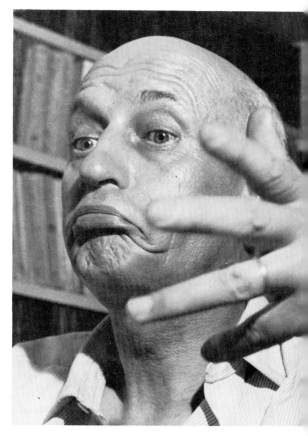

THE BEARD

■ The beard is the most conspicuous of all human gender signals, visible at a great distance and in almost any state of dress or undress. Man may not be the only species of primate to sport a beard, but he is certainly the one with the most impressive appendage. The world-record length for a beard is 17½ feet, and for a moustache the maximum span recorded is 8½ feet. Although these figures represent rare extremes, any normal adult male human who ceased to trim his beard and moustache for a decade would make an awesome impact as a hairy-faced primate, easily outclassing the various other mustachioed and bearded species. The combination of bushy facial hair and flowing scalp hair in the completely unbarbered prehistoric male must have been an amazing sight to behold and something quite novel on the animal scene.

The growth of human facial hair begins slowly at puberty, under the influence of male hormones. The typical adult female never sprouts more than a very fine 'peach-fuzz', which is invisible at a distance and can only be detected by unusually close scrutiny. By contrast, the adult male grows long hairs around his mouth, all over his lower face, his jaws and chin, and his upper throat. These hairs increase in length by about a sixtieth of an inch per day, which means that if he stopped shaving completely he would be the proud possessor of a foot-long beard in just over two years. This would cover much of his chest region and no other biological feature would make him look more different from the adult female.

The primary function of the male beard has been hotly debated for many centuries. A strong case was made for it as a natural 'scarf' keeping the delicate throat region safe and warm. It was claimed that because the males had to go out on the hunt, exposed to all weathers, while the females and young stayed snugly at home in the family cave, nature had provided the brave men of the tribe with an exclusive form of protection. There are two major flaws in this theory. Firstly, if the bearded males had developed clothing of some kind – animal skins, for instance – to protect the rest of their (functionally hairless) bodies, they could easily have added some sort of throat protector, if this was so important to them. If the beard is envisaged as developing before the advent of clothing then, of course, there is no sense in leaving much of the body naked while protecting the throat. If rugged, hunting males, out on the chase on a cold, ice-age morning, were in need of hairy insulation, nature would surely have given them back their whole pelt of fur. Secondly, those human races most perfectly adapted to the colder regions of the world – the fat-lined eskimos, for example – happen to display the least-bushy beards. If beards were throat-warmers, they should be the hairiest of all. The alternative theory sees the male beard as nothing more than a display of masculine adulthood, a simple gender signal with no other properties.

As a signal of masculinity, a kind of male flag, the beard's main impact is strongly visual, but it also appears to operate as a scent-carrier. The facial region possesses a number of scent glands and their products are retained better on a hairy face. During adolescence, when these glands are being brought into play for the first time, an excess of hormones can cause the skin disturbance we refer to as acne. It is a cruel twist of fate that the sexiest of teenagers are those with the most severe acne rashes.

■ It takes just over two years for the average man to sprout a foot-long beard. Generous and impressive beards of the kind seen here are therefore the result of five or six years growth, and a great deal of careful grooming. No other species of animal can boast a chin appendage approaching this length. The shaved masculine jaw can also make an impression, especially if coupled with an assertive posture. The aggressive male juts his chin forward, and it is no accident that, in our species, males have more-prominent chins than females.

Not only is the beard the male gender flag – it is also a sexual scent carrier.

■ *Like many of their primate relatives human males possess facial hairtufts which act as gender signals. Untrimmed, the human beard grows to dramatic proportions and helps to emphasize the aggressively jutting chin posture of the assertive, competitive male. A powerful beard display automatically makes a male look less friendly and more threatening, which today is the main reason for the removal of the beard by shaving.*

As a visual signal denoting a dominant adult male, the beard helps to exaggerate the aggressive human posture of the jutting jaw. When we are angry, we project the chin forwards; when we are submissive, we pull it in and back. Males have heavier jaws and more protuberant chins than females, giving them a more aggressive facial shape, even in moments of complete relaxation. With beards added to this projecting display, they become even more juttingly hostile in their appearance. Aiding this is the fact that beard hair is more kinked and curly than scalp hair, so the beard has more body and stands out thicker on the face.

It is because of this difference that hero figures are always portrayed with massive jaws and impressively proud chins. Unfortunate males with receding chins are insultingly dubbed 'chinless wonders' and looked upon as effeminate 'wets'. Conversely, strong-jawed women are automatically regarded as tough, rugged personalities. These responses persist regardless of conflicting evidence about the true characters of known individuals, because we cannot help reacting, at an unconscious level, to these ancient biological gender signals.

Given its importance as a masculine signal, the removal of the male beard by daily shaving appears both bizarre and perverse. Why

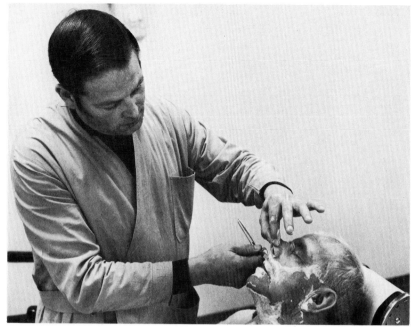

should so many adult males in so many cultures have wanted to throw away their most dramatic gender signal? It is not as though it is an easy matter, for shaving takes time and trouble. A sixty-year-old man who has spent ten minutes a day removing his stubble will have wasted no fewer than 2,555 hours – or 106 days – on this strange task, assuming that he began to shave regularly at the age of 18. Why does he do it?

In ancient times, the beard was looked upon as the male symbol of power, strength and virility. To lose one's beard was a desperate tragedy. It was the punishment meted out to a defeated foe, to prisoners and to slaves. To be clean shaven was a disgrace. Men swore on their beards; beards were sacred; God was heavily bearded, and a shaven deity was unthinkable. The pharoahs of ancient Egypt always wore false beards for ceremonial occasions, to demonstrate their high status and masculine wisdom. Even the female pharoah Queen Hatshepsut wore a false beard to display her great power.

Although today bearded ladies are thought of strictly as circus freaks, in ancient times certain of the mythological Mother Goddess figures were shown as bearded in order to give them greater significance. Even the Christian church was able to boast a bearded female martyr – Saint Wilgefortis, a virgin who met her death by crucifixion.

Because beards were so important, the rulers of early civilizations such as Persia, Sumer, Assyria and Babylon devoted an immense amount of time to grooming and decorating them. They used tongs, curling-irons, dyes and perfumes. Their beards were coloured, oiled, scented, pleated, curled, frizzled or starched and, for special occasions, were sprinkled with gold dust and shot through with gold thread.

The earliest examples of voluntary shaving appear to have been connected with the desire to display an 'enslavement to a god'. Young men would offer their beards to their deities as a sign of loyal submission. Priests might shave off their beards as a symbol of humility. But shaving on a more permanent and widespread basis seems to have been introduced as a military style in ancient Greece and Rome. Alexander the Great is reputed to have instructed his troops to remove their beards in order to improve their chances in close combat. It was considered that long beards would make useful handholds for the enemy – as one sometimes sees today in professional wrestling bouts. Roman soldiers were told to shave their beards for reasons of identification. Their shaven chins were easily distinguished from those of the hairy barbarians they were fighting. There was also the

▲ For many years bearded ladies were a major attraction at circus sideshows; but with changing public taste and the gradual decline of freak shows, they have become a great rarity. New depilatories make it easy for a woman with a hirsute chin to remove the offending hair, and there is little point in keeping it if there is no longer a career to be made out of it. The strange biological feature of bearded ladies is that they do not appear to be generally more hairy over their whole bodies. The most famous ones from Victorian times were reported to have 'smooth and beautiful skin' apart from that on their chins. The genetic factor that causes the bearded female face appears to be highly specific.

■ *Because a frightened man tucks his chin in and an aggressive one sticks it out, it follows that any individual who has a naturally receding chin will find it hard to assert himself and will be scathingly referred to as a 'chinless wonder'. The man with a bigger chin will be described, admiringly, as square-jawed. These anatomical variations have little to do with true assertiveness. Many a man with a weak chin – such as Frederick the Great, for example – has proved to be highly assertive in real life. Individuals with unusually protruding jaws often display a small cleft or dimple in the centre of the chin, a feature which is widely considered to be appealingly masculine.*

■ *The fact that a shaven man looks more feminine has sometimes aroused the scorn of the bearded. Writing in the seventeenth century, John Bulwer ranted and raged against the naked chin: 'What greater evidence can be given of effeminacy than to be transformed into the appearance of a woman, and to be seen with a smooth skin like a woman, a shameful metamorphosis...How more ignominious it is in smoothness of face to resemble that impotent sex?' Shaving, he roared, was 'a ridiculous fashion to be looked upon with scoffs and noted with infamy...The beard is a singular gift of God, which who shaves away, he aims at nothing than to become less man. An act not only of indecency, but of injustice, and ingratitude against God and Nature, repugnant to Scriptures...' He continues in this vein for some twenty-four pages.*

question of improved hygiene, although it is not certain how big a role this played.

Once there were two well-established alternative fashions, shaven and unshaven, it became possible for any social group or culture to make a statement of allegiance or rebellion by the way their males trimmed their beards. Sometimes the pendulum swung in a new direction simply because of a leader's shaving habits. One French king had an ugly scar on his chin and grew a beard to hide it. Out of respect all Frenchmen of the period wore beards. One Spanish king was unable to grow a beard, so all Spaniards of the period went clean-shaven to honour him.

Sometimes, one part of society wore beards while others went clean-shaven, giving us the 'bearded warrior', the 'bearded sailor', the 'bearded artist' and 'the bearded hippie' as special types within a predominantly well-shaved population. When this happened, the bushy exceptions to the general rule were usually either aggressively dominant males, or wild, shaggy ones. The aggressive military beards were typically trimmed and pointed, indicating that their owners were both domineering and well organized. The shaggy hair of the social rebels, such as painters, poets and hippies was more likely to be unruly and straggling, reflecting their owners' lack of regard for social conventions and controls.

Despite these special categories, the ordinary 'mainstream male' has nearly always been clean-shaven in recent centuries. Occasionally this has been imposed on the population, as in Elizabethan times in England when a tax was levied on beard-wearers, the lowest rate being the then considerable sum of three shillings and fourpence, per annum. This restricted beards to the upper classes and made them into a financial status-display. In other cases, severe social ostracism was the fate of the bearded, which only the most stubborn and courageous could resist. In Massachusetts in 1830, a bearded man in one town had his windows broken, stones thrown at him by children, and communion refused by the local church. Eventually he was attacked by four men who tried to shave him by force. When he fought back, he was arrested and jailed for a year for assault.

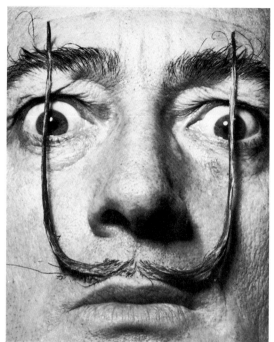

■ *The moustache has often been referred to as a symbol of obsessive but inhibited sexuality and masculinity. It has also been bluntly stated that a man with a moustache but no beard is a man with sexual problems. The basis for these remarks is simple enough. The wearing of a moustache reflects a need to demonstrate masculine gender, but the restriction of facial hair to this small, carefully trimmed zone on the upper lip indicates restraint and tight self-control. There is an element of truth in this interpretation, but applied as a general rule it becomes meaningless, as is obvious from the variety of examples shown here. For many men the decision to wear a moustache is not a personal one but simply a response to a dominant local fashion.*

Usually, such strict measures were quite unnecessary, the majority of men being happy to shave without any encouragement. Only in cultures which were temporarily dominated by military hostilities, or by a pompous and patriarchal attitude towards the male role (as in late-Victorian society) did the beard become widespread and flourish for a brief spell as the 'social norm'.

This may give us a clue in answer to the earlier question: Why do men waste time every morning shaving the stubble from their chins? If beards display male hostility and dominance, then the removal of beards on a voluntary and regular basis must indicate a desire on the part of men to damp down their primeval assertiveness. In other words, the clean-shaven male is making a visual statement which is a request for cooperation rather than competition. He is saying to his companion: 'I am reducing my level of masculinity in the hope that you will reduce yours. In this way we can work and play together without over-stimulating our ancient pugnacity.'

Viewed in this way, shaving becomes a worldwide appeasement display. It also has several side-benefits. Since children do not have beards, it makes the shaven male look younger than he is, by unconscious association. It also reveals the facial expressions more clearly and therefore makes the shaven male more communicative. A really heavy beard tends to obscure friendly expressions, which suits a domineering male, but not 'one of the boys'. It also displays personal grooming care and self-attention, which implies cleanliness. A bearded eater is potentially a more messy eater than a 'clean-shaven' one.

So shaving makes an adult male look more juvenile, more friendly and cleaner. But it also makes him look more feminine – and this has sometimes aroused the scorn of stern bearded critics. Although most people ignore this effect of shaving, there are some who privately feel that it is more 'macho' to be hairier. But they have a problem. They want to look young and expressive and 'clean-cut', but they would also like to display a little masculine hair to flaunt their gender in a modest way. The answer is that famous compromise, the mous-

▲ *Stroking one's chin is a relic gesture dating back to the time when the beard was a traditional symbol of wisdom and when moving a hand through it was meant to indicate deep thought. The beard as a symbol of slow growth and the passage of time is also exploited in many regional signals. In Germany and Austria, for example, the thumb and forefinger of the same hand are sometimes stroked down either side of a real or imaginary beard in a gesture which means: 'The joke you are telling me is so old that it has grown a beard.' In France and northern Italy, for example the back of the hand may be flicked forward against the underside of the chin, as if rubbing against the beard. This is a gesture of boredom, carrying the message: 'Look how my beard has grown while I have had to endure this.' Done aggressively, however, this chin-flick gesture exploits a third symbolic property of the beard and means: 'I point my masculinity at you!' It is performed as an insult towards someone who is thought to be lying or is making a nuisance of himself. It is a modern gestural version of the primeval male beard-threat display.*

■ *Gestures involving the touching of the chin of another person are comparatively rare. Women may gently stroke the beards of their male lovers from time to time, but this is no more than part of a general fondling. The only specific form of jaw touching is the 'chin chuck' in which a man pushes the side of a forefinger up under the jawline of a woman. The message is 'keep your chin up', or 'keep cheerful', but the intrusive intimacy of the action converts it into a slightly bossy demand, even though made with love. For the male in question the action also serves to remind him, through the skin of his finger, of the feminine smoothness of the woman's face.*

tache. By wearing only the part of the beard that covers the strip of skin between the nose and the upper lip, they can have the best of both worlds. A mustachioed figure is clearly not female, and yet it retains the expressiveness and the cleanliness of the shaved face. On the juvenility scale, it looks mature but not old.

The macho moustache has been popular with many military men, who have waxed it, dyed it, preened it and twiddled it, often making it into the focal point of their masculinity. Each period has seen its own special style of moustache-shape, from the handlebars of the war-time RAF pilots to the tiny, pencil-slim strips of the early movie stars. As a symbolic statement, the moustache says: 'I want to be considered friendly (which is why I have shaved my chin) but I also think of myself as exceptionally masculine (which is why I have left a reminder of my great male beard).'

Sometimes there are sudden shifts in the social significance of a moustache. A classic example of this has taken place recently in New

York and San Francisco, where macho males have lost their moustache signals to the gays. When the homosexual males started to wear moustaches in the 1970s, the old-style, tough heterosexuals who also wore them found themselves being followed down the streets by eager would-be gay partners. Horrified, they quickly shaved off their ex-macho badges and a new phase of moustache display was born in these cities.

Finally, there are two moustache gestures – the Moustache Wipe and the Moustach-tip Twiddle – which both have the same meaning. They are preening actions implying 'preparation for courtship' and are used as signals of arousal when an Italian male sees an attractive female and wishes to convey his feelings about her to his friends. The twiddling gesture dates from the days when waxed mustachios were commonplace and it has outlived them. Today, clean-shaven men will twist the imaginary tips of non-existent mustachios, but the significance of the gesture is not lost on their companions, even though they may be too young ever to have seen a waxed moustache.

THE NECK

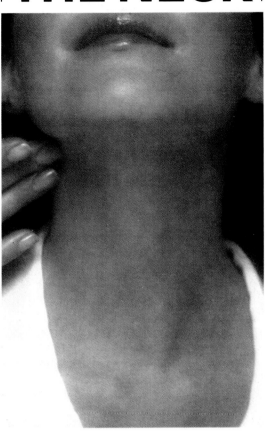

■ The neck has been described as the most subtle part of the human body. Besides containing the vital connexions between mouth and stomach, nose and lungs, and brain and spine, it houses the crucial blood vessels between heart and brain. And surrounding these connexion lines are complex groups of muscles which enable the human head to dip and nod, shake and twist, turn and toss and perform a whole range of movements that convey important messages during social interactions.

Traditionally, the exceptionally masculine figure is 'bull-necked' while the exceptionally feminine figure is endowed with a graceful 'swan-like' neck. These differences are real enough. The male neck is shorter and more thickset, the female neck longer, more slender and more tapered. This is partly because of the stronger musculature of the male and partly because the female has a shorter thorax, the top of her breastbone being lower in relation to the backbone than the male's. This gender difference undoubtedly developed during the long hunting phase of human evolution, when males with stronger, less-snappable necks were at an advantage.

Another gender difference in the neck concerns the Adam's apple, which is much more conspicuous than Eve's equivalent. This is because women, with higher pitched voices, have shorter vocal cords requiring a smaller voice-box. Male vocal cords are about 18 millimetres long, those of the female only 13. The male larynx is roughly 30 per cent larger than that of the female. It is also placed slightly lower down in the throat, which has the effect of making it even more prominent. This laryngeal gender difference does not appear until puberty, when the male voice deepens, or 'breaks'. The adult female voice and the female larynx are essentially more infantile than the male's, retaining a pitch of between 230 and 255 cycles per second, while the adult male voice sinks to somewhere between 130 and 145 cycles per second. For some inexplicable reason the voices of adult males living in small tribal societies are higher pitched and more shrill than those of males in modern urban societies. Equally mystifying is the observation that professional prostitutes have a larger larynx and deeper voice than other women. Why their occupation should make them vocally more masculine is not clear, although it suggests that their unusual sexual lifestyle may in some way disturb their hormonal balance.

The origin of the term 'Adam's Apple' is not hard to guess. Early folklore maintained that the lump in the male throat was placed there to act as a constant reminder of Adam's original sin – eating the apple offered to him by Eve. A piece of this apple was said to have become immovably lodged in his throat when he took his first bite of the forbidden fruit. In actual fact the word 'apple' is not used in the Bible story of the Garden of Eden. It is a later invention, but the name 'Adam's apple' has defiantly survived.

Because the female neck is more slender than the male's, artists have frequently exaggerated this feature to create super-feminine images. Cartoonists portraying attractive women invariably narrow and lengthen the neck even further than normal anatomy would permit. Also, model agencies, when selecting individual girls for training, choose those with necks that are thinner and longer than the average. In one culture this desire for long-necked women was taken

■ The neck, with its complex groups of muscles, is responsible not only for the entire repertoire of human head signals but also for the quality of body posture.

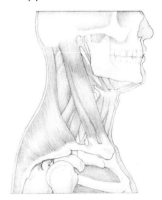

■ *The female neck is longer and more slender than that of the male. Even male ballet dancers, who develop the verticality of their postures to an extreme, cannot compete with females for neck length. This is partly because there is an anatomical difference in the length of the thorax, which is shorter in the female. Many men, as they grow older, hunch up their shoulders so that the neck becomes thick and squat, making the gender difference even more extreme.*

to bizarre extremes. The Padaung branch of the Karen people of upland Burma boasted what came to be known in Europe as 'giraffe-necked' women. The word 'padaung' means 'brass wearer' and the females of this group were required by local fashion to start wearing brass neck-rings from an early age. To begin with, five rings were fixed around the neck and this number was increased gradually, year by year, until it reached a total of 22 or even, in some cases, 24. Brass rings were also put on the arms and legs, so an adult female might carry around with her a weight of brass totalling between 50 and 60 pounds. Despite this encumbrance the women of the tribe were expected to walk long distances and work in the fields.

The most amazing aspect of this custom was the extent to which it artificially lengthened the women's necks. The record neck-length recorded was 15¾ inches. The cervical muscles were stretched so severely by this practice that the neck vertebrae were pulled apart in a completely abnormal way. It was said that if such a woman had her heavy brass rings removed her neck would be unable to support her head. Europeans, fascinated by this horrific cultural distortion of the human body, paraded a number of these long-necked women in circus side-shows – until human freaks were no longer considered to be objects of casual curiosity.

In occult circles the neck has always been a body zone of major importance. In some cults, such as voodoo in Haiti, it is believed that the human soul resides in the nape of the neck. It was the occult significance of the neck which led to the widespread use of necklaces in earlier days. They were more than mere decorations, having the special function of protecting this vital part of the human anatomy from hostile influences such as the 'evil eye'.

The neck became the focus of certain occult ritual practices. It was discovered that by applying pressure to the large carotid arteries

Neck posture and neck presentation.

which run up the side of the neck, carrying blood to the brain, a subject could be made dizzy and confused – an easy prey to suggestion. What was happening, of course, was that the subject's brain was being deprived of oxygen; but in the mumbo-jumbo of religious rites, his condition could conveniently be attributed to the supernatural.

A much healthier form of neck manipulation was developed by Matthias Alexander, who founded what was to become known as the Alexander Treatment. This was based on the idea that by modifying the basic posture of the neck on the shoulders, it was possible to cure not only certain physical symptoms but also a variety of psychological disturbances. Some critics have argued that this concept gives the neck an almost mystical power over the rest of the body, but there is a simpler explanation. Because urban man spends so much time hunched over a desk or a table or slouched in a chair his neck gradually loses its natural vertical posture. If through Alexander training this posture can be re-established, the rest of the body automatically follows suit and recovers its correct balance: the scene is set for a return to a healthy body tonus, which may in turn lead to a healthier mental state. It is really no more mystical than the kind of posture training a ballet dancer receives. In both cases the neck seems to be the key that unlocks the body's poise.

Turning to gestures, there are comparatively few that focus specifically on the neck. The most widespread is the Throat-cut Mime, in which the gesturer uses his hand as a mock knife to slice across the front of his throat. This has three closely related meanings. If done in anger it can indicate what the gesturer would like to do to someone else. If done as an apology it shows what the gesturer feels like doing to himself. If performed in a television studio by the studio manager it is a silent signal to the presenter of the programme telling him that time has run out. The three messages, respectively, are: 'You will die'; 'I could cut my throat'; and 'If you don't stop now you'll be cut off'. Also widespread is the 'mock self-strangling' gesture in which the hand or hands clasp the gesturer's neck and pretend to throttle it. As with the mock throat-cut this may signify either 'I want to strangle you' or 'I could strangle myself.'

Another popular gesture is the 'I-am-fed-up-to-here' signal, in which the forefinger-edge of a palm-down hand is tapped several times against the Adam's apple. The implication is that the gesturer has been stuffed so full of something that he cannot take any more.

More important than these regional gestures are the many neck actions which result in head movements or postures. These are of two kinds. First, there are actions which adjust the head to the environment, as when we turn the head to look at something, cock the head to listen to a sound, raise the head to sniff the air, or lower the head to shovel food into the mouth. Second, there are the actions whose sole function is to transmit visual signals to our companions. There are more than twenty of this second kind:

1. The Head Nod. The neck moves the head vertically up and down one or more times, with the up elements and the down elements of approximately equal strength or with the down elements slightly stronger. This is the most common and widespread action of assent or agreement. Although in some cultures there are other actions signifying assent, it remains true that whenever it does occur it

▲ *This Japanese Geisha is applying a traditional form of neck make-up to enhance her beauty. The back of the neck is an important erotic zone in Japan.*

In the image: handwritten German-style text including "hrliche alten", "Sortilana in foll die foll", "ihrer haben", "und die voll", "O tempora ô mores".

▲ *These elaborate neck adornments of 1630, aptly described as 'millstones around the neck', had the effect not only of incapacitating the wearers but also of making them appear decapitated – a fate which befell some of them in a less symbolic form.*

▼ *For thousands of years women have been tempted to wear some form of neck decoration on special occasions, which makes this young woman's naked neck – at her coming-out party – unusual. The absence of a necklace or neck band helps to draw attention to the beauty of the strongly feminine, tapering shape of her neck.*

has the same confirmatory meaning: a Head Nod always means 'yes'.

Nodding has a global distribution. Early travellers found it in remote tribal societies which had not previously encountered western influences. In cultures where some other signal was used for 'yes' this was usually an alternative to rather than a replacement for, the Head Nod. In parts of Sri Lanka, for example, the Head Nod is used when responding to a factual question, but another movement, the Head Sway, is made when agreeing to a proposal. Both say 'Yes', but they are a different kind of 'Yes'. Some societies make subtle distinctions between one kind of affirmation and another while others simply lump them together and give a blanket Head Nod in every situation calling for assent. One thing is clear: if two observers in Sri Lanka had asked two slightly different questions they might have come away with conflicting views regarding the local affirmation signal. In the past this has happened all too often.

Two suggestions have been made about the origin of the Head Nod. The first sees it as a modified version of the bow and interprets it as a highly-abbreviated submissive body-lowering device. By saying 'Yes' one is in a sense momentarily submitting to the other person. Viewed in this way the Head Nod could be described as a quickly checked 'one-per-cent prostration'. The second suggestion relates the Head Nod to the infantile act of feeding at the breast. The nodding movement is seen as part of the pattern of accepting the breast. Rejecting the breast is characterized by the baby jerking its head sideways or upwards – and on this basis the negative head signals of adults should involve sideways or upward movements. As we shall see, there is some truth in this.

2. The Head Bow. The head is lowered into a head-down posture and then raised again. This action differs from the Head Nod in several details. Unlike the nod it is always a single down and up action and the movements are stiffer and more deliberate. Also, at high intensity the head may be held momentarily in the down position before returning to neutral, which never occurs in the nod. Finally, it does not carry the head above the neutral line, as does the typical nod during its up movements. The Head Bow appears to be an almost worldwide greeting signal. In origin it is clearly a low intensity version of the general body-lowering tendency of a submissive individual. However, although its basic message is 'I lower myself before you', it is not restricted to subordinates. When performed by equals or by dominant individuals the message is expressed in its negative form as 'I am not going to assert myself', which becomes generalized as 'I am friendly'. In strength the action varies from an almost imperceptible bob of the head, to a dramatically exaggerated snap of the neck. The main cultural difference appears to be in the stiffness of the action, the Oriental Head Bow being much softer than the 'Germanic' version. Besides its main function as a greeting signal, the Head Bow also appears as a parting signal and at moments when gratitude is being expressed.

3. The Head Dip. The neck jerks the head forcefully downwards and then returns it to neutral. This action is similar to the Head Bow but the downward movement is much more vigorous, or, to be more precise, there is a much greater difference between the strength of the down element and the up element. The Head Dip is used to em-

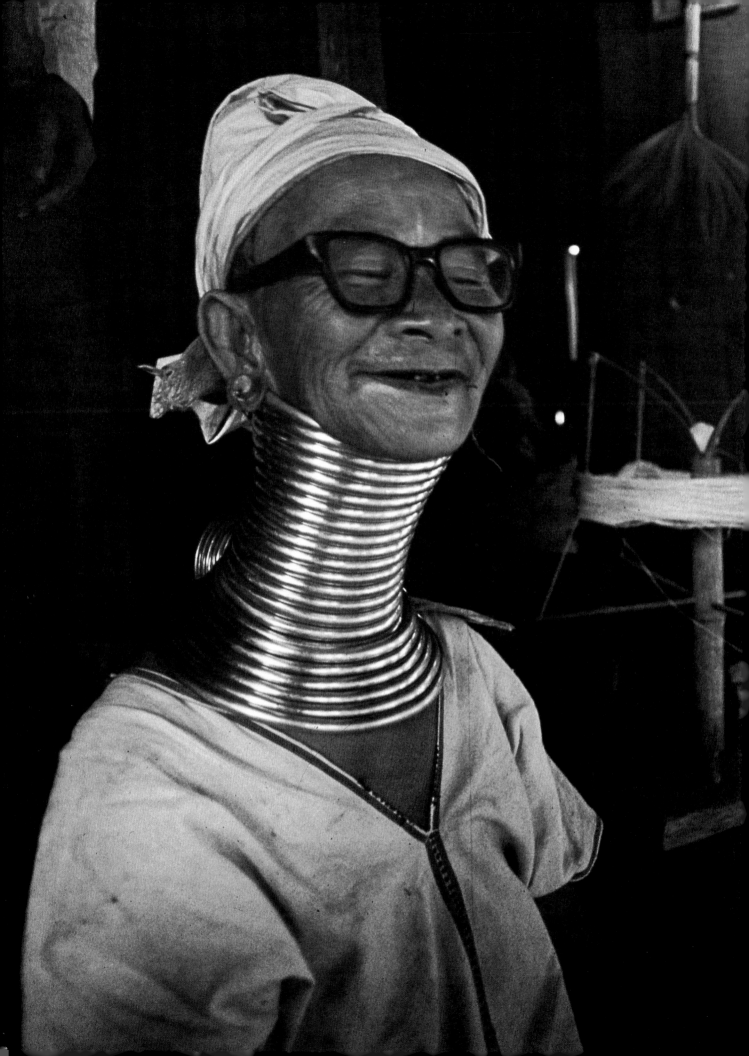

◄ *Because females have longer necks than males, it follows that to exaggerate this difference by stretching the neck upwards is to increase the femininity of the individual. The most bizarre example of this trend is to be found among the 'giraffe-necked' women of Burma. Starting as young girls, they have more and more brass rings fixed around their necks until their neck bones start to pull apart. They are still able to move about and even work in the fields in this condition, but if their rings were ever removed it would probably kill them because their necks would no longer be able to support the weight of their heads.*

▼ *Although this Kenyan girl has collected a huge number of heavy, brightly coloured necklaces, which seem to be thrusting her head upwards and severely stretching her neck, she is not suffering from neck distortion like the Burmese women. If the necklaces are removed, her neck still functions normally, but while she is wearing them they force her to adopt a feminine, tall-necked posture.*

phasize the spoken word. Speakers who are engaged in assertive or aggressive conversations and wish to drive home the points they are making do so with repeated dips of the head. It is almost as if, with each new dip, the head is being thumped down like a symbolic fist on an imaginary table. There is a slight forward movement as the head jerks down, giving it an attacking quality that underlines the forcefulness of the accompanying word. The Head Dip also occasionally occurs without speech, when the silent message is 'So there!'

4. The Head Toss. The head tilts sharply back and then returns to the neutral posture. This action is the opposite of the Head Bow, the head going up instead of down. It is used in several quite different ways, which sometimes leads to confusion. Its most widespread occurrence is as a distant friendly greeting, performed right at the beginning of the encounter before closer interaction has taken place. Its message is 'I am pleasantly surprised to see you'. Surprise is the key factor here, the Head Toss representing a highly modified and reduced element of the full startle response. That the Head Toss and its opposite the Head Bow can both be used as greeting gestures is explained by the fact that the toss occurs at a distance and the bow at close quarters. Also, the toss is a gesture of familiarity, the bow of formality.

A second use of the Head Toss is as a signal of understanding. In this role it occurs during a conversation at the moment when someone suddenly sees the point of something and exclaims 'Ah yes, of course!' Again this is a moment of surprise – it is as though the person concerned has suddenly started up or jerked backwards in a fleeting moment of alarm. The surprise is quickly dissipated and the head returns to neutral, leaving only the rapid Head Toss as a tiny intention movement of retreat.

A third use borrows the 'Ah-yes' element from the last example and turns the Head Toss into another way of saying 'Yes'. This is said to occur in a few corners of the world such as Ethiopia, the Philippines, and parts of Borneo. It is not a common form of the movement however, presumably because it clashes with the Head Nod 'yes'.

A fourth version signals the exact opposite: 'No!' This is popular in Greece and neighbouring countries, but again it is not widespread, this time because it clashes with a much more popular negative signal, the Head Shake. The so-called 'Greek No' puzzles many visitors to the Eastern Mediterranean when they first encounter it. If they ask a civil question and in reply receive the sharp upward flick of the head, which is the main constituent of the Greek Head Toss, they imagine that they have irritated the gesturer, but cannot understand why. The reason for this is that in other parts of Europe there is a common 'irritation reaction', widely understood, in which the head jerks upwards, the eyes gaze upwards and the tongue is clicked. This transmits the message 'How stupid!' and the 'Greek-No' gesture looks so similar that to a visitor it seems like a criticism of his enquiry. But to the Greek it simply means 'No' and carries no impoliteness.

In origin the Greek negative Head Toss appears to derive directly from the ancient breast-rejection reaction of the baby. Attempts by parents to spoon-feed an infant who is not hungry can easily produce a similar upward flick of the head and it is easy to see how this could grow into an adult negative gesture.

5. The Head Shake. The neck turns the head from side to side, with

There is more to conversation than the talk.

▲ *The twisting of the neck in the direction of someone else is a particularly human way of demonstrating cooperative attention and social engagement. The failure to make such movements when involved in conversation is a harsh display of aloofness and uninterest. Individuals who are deliberately ignoring those around them often display their mood by stiffening their necks and refusing to turn their heads towards their companions.*

▶ *The complex muscles of the human neck give us considerable mobility of our heads. We can arch, crane, stretch, twist and dip our necks in many different ways, enabling us to stare in almost any direction without necessarily moving the rest of the body.*

equal emphasis left and right. This also originates in infantile food refusal, either at the breast or when being spoon fed. As a result the Head Shake is essentially a negative signal whenever it occurs. It transmits messages ranging from 'Certainly not!' to 'No thank you' to 'I don't believe it.' It has been suggested that in certain countries the Head Shake signifies 'Yes', but this is based on poor observation. The lateral head movement which sometimes indicates an affirmative is of a different type and is discussed below as The Head Sway. So basic is the Head Shake negative that other parts of the body may be used to mimic it. The Forefinger Wag and the Hand Wag are both gestures which imitate the shaking head, transmitting the same negative signals.

6. The Head Twist. The neck pulls the head sharply to one side and then back again into the neutral position. This is a one-sided head shake and is even more like an infant's food-refusal movement. It is a rare variant of the shake which carries the same 'No!' message in Ethiopia, for example.

7. The Head Side-jerk. The head half-turns and half-tilts to one side, the two short movements occurring simultaneously. It is difficult to describe this small but complex movement, but it could be called a 'winkless wink'. When most people perform a heavy wink they augment the eye-closing element with an accompanying head movement. This exaggeration has become isolated and emancipated from its original pattern, so that it can now be given on its own without the actual wink. This Head-Side-Jerk becomes a friendly mock-conspiratorial greeting. It is employed at fairly close quarters. Sometimes it is used in television commercials more as a comment than a greeting. The actor sipping the drink grins and performs the side-jerk, conveying the message that 'Between you and me, this is rather good'. Conspiratorially he lets us in on a good thing.

8. The Head Beckon. The neck pulls the head away from the companion, jerking it backwards with a slightly tilted posture. This gesture says 'Come with me' or 'Come over here' and is a substitute for the hand or forefinger beckon. It is most likely to occur when the hands are full or the beckoner wishes to signal without being too conspicuous. Today it is sometimes used as a joke invitation to sex, no doubt gaining this flavour from its secretive quality. Interestingly, when the companion is directly in front of the head-beckoner the latter still adds a slight tilt to the jerk of the head. In this way the message remains distinct from that of the Head Toss.

9. The Head Wobble. The neck moves the head from side to side as in the Head Shake but with much smaller, shorter, quicker movements. Performed by an older man, as it usually is, it tends to make his jowls quiver and wobble from side to side. This is a curious action commonly seen among dominant males who hold some position of authority – politicians, generals, administrators. It always accompanies speech and may coincide with a verbal emphasis that is either positive or negative. The head may start to wobble laterally as the speaker intones either 'There is every hope' or 'There is no hope'. In the second case it is easy to understand why the speaker may wish to perform a suppressed Head Shake, but it is not so easy to say why the Head Wobble also occurs with positive statements. Possibly the speaker is lying and is trying to suppress a truthful negative Head Shake but failing to do so, leaving the wobble as a tell-tale clue.

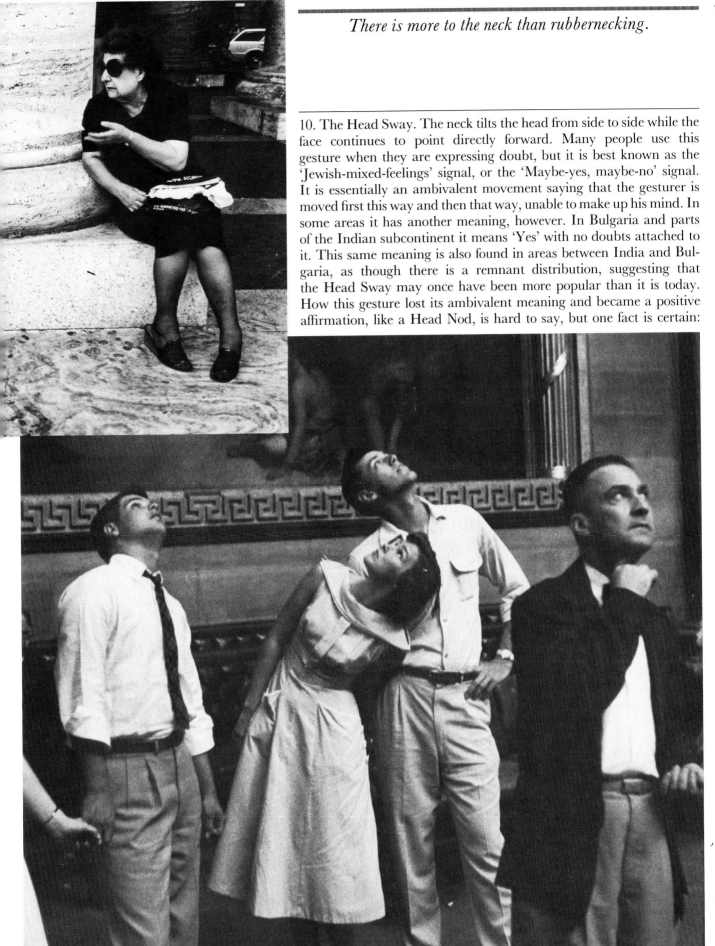

10. The Head Sway. The neck tilts the head from side to side while the face continues to point directly forward. Many people use this gesture when they are expressing doubt, but it is best known as the 'Jewish-mixed-feelings' signal, or the 'Maybe-yes, maybe-no' signal. It is essentially an ambivalent movement saying that the gesturer is moved first this way and then that way, unable to make up his mind. In some areas it has another meaning, however. In Bulgaria and parts of the Indian subcontinent it means 'Yes' with no doubts attached to it. This same meaning is also found in areas between India and Bulgaria, as though there is a remnant distribution, suggesting that the Head Sway may once have been more popular than it is today. How this gesture lost its ambivalent meaning and became a positive affirmation, like a Head Nod, is hard to say, but one fact is certain:

The bowed head can also signify a temporary withdrawal from harsh reality.

◄ Many neck movements result in head postures or actions which carry specific mood-messages. The lowered head is the typical posture of defeat and dejection, commonly seen in sportsmen who have just lost an important contest. The lowering of the head not only reduces the height of the defeated individual, but it also effectively cuts out incoming stimuli, stimuli from a world that is suddenly unappealing.

earlier travellers who observed it and reported that it was a Head Shake meaning 'Yes' were mistaken. They were confused because of its lateral elements, but the two movements are quite distinct.

11. The Head Point. The neck points the head to give a direction, jerking it towards the person or object of interest. The human animal is well equipped for precise pointing with forefinger or hand, but occasionally the hands are busy or local rules forbid hand-pointing. When this happens the head takes over the role of pointer.

12. The Head Jolt. The head performs a single short sharp shake. In origin this is the action of clearing the head, but it is often used as a deliberate signal to indicate 'mock-surprise' or shock. The implication is that the information just received is so extraordinary that the Head jolt is necessary to clear the head and make sure the listener is not dreaming.

13. The Head Freeze. The head is held rigidly in its neutral position. This is a non-gesture which becomes a gesture by being inappropriate. For example, someone picks up a vase and smashes it. Everyone present quickly turns to look at what has happened. If one person there freezes his head and refused to turn it or move it in any way, this says a great deal about his reaction. The deliberate absence of movement, when it is appropriate to respond, suggests either that he is so dominant and fearless that even in close proximity to something being smashed he will not bother to look at it, or that he is bored because he has seen this kind of thing before and feels it is beneath his dignity to stare at it again.

14. The Head Slow-Turn. The neck turns the head sideways very slowly towards a point of interest. When the movement is extremely slow and the stimulus is strong, the message is similar to that of the Head Freeze. It is inappropriately calm and therefore suggests, again, either dominance or boredom or both.

▲ The bowed head, like the lowered head, is a submissive, subordinate gesture, but it has become formalized as a brief down-and-up movement, rather than a prolonged 'down' posture. As a form of body-lowering it can be amplified by a crouching body posture, as in the case of this Geisha from Kyoto. She is entering a room where a male visitor is awaiting her and she makes him feel superior by deliberately adopting an inferior, head-bowed position.

► This African Pygmy woman is adopting a shy head-cock posture in which the neck tilts the head to one side, as if to lay it on the left shoulder, which is slightly raised, as if to receive it.

▲ The neck is such an important erogenous zone that the expression 'necking' was invented to describe preliminary sexual contacts between young lovers. In some cultures – American and Japanese, for example – a special kind of erotic neck-biting is popular. Its importance is that it leaves small red marks on the neck which take some days to fade and which therefore act as a telltale sign of sexual activity. In reality, the red marks are not made by a true bite, but by a prolonged suck, the partner pressing the lips to the delicate neck-skin and then sucking hard for several seconds before letting go.

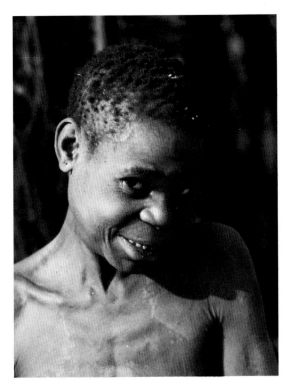

15. The Head Aside. The neck turns the head sideways away from the point of interest. This is basically a protective device, either turning the face away from physical threat or removing the sense organs from sights or smells they wish to avoid. In special cases it may be used as a way of concealing identity by hiding the face. But as a specific, deliberate gesture it is used as a 'cut' or rejection. In this form it is a silent insult telling someone that you are disengaging from social contact with them. But unless the movement is done well there is the problem that the studied insult might be mistaken for a shy hiding of the face. It requires that the insult movement be done boldly and exaggeratedly. Unfortunately this gives it a somewhat pompous flavour which has tended to curtail its use in modern times, although it was extremely popular in the last century. Its main use then was to keep the nouveaux riches in their place. The industrial revolution had dramatically increased the wealth of the middle classes whose dearest wish was to use their new position to elevate themselves in social rank. This the upper classes resisted by introducing 'the cut'. Etiquette books explained the technique: 'The individual should be allowed to see that his approach has been noticed, whereupon you turn your head away.' Today although the formal cut is extremely rare it is still common at moments of intense petulance during family quarrels.

16. The Head Advance. The neck thrusts the head forward towards the source of interest. This can be done with love or with hate. Lovers crane their necks forward to gaze gently into each other's eyes. Haters jut their heads forward to display their fearlessness and glare into each other's eyes. A third context in which this action occurs is when someone is so desperate to claim your entire attention that he or she thrusts his or her face right in front of yours in order to prevent you from seeing anything else that might distract you.

17. The Head Retreat. The head is withdrawn from the source of interest. This is basically a simple avoidance action, but it may also be used as a deliberate signal. If given in response to a comment, the gesturer is in effect saying 'I want to distance myself from that statement' and does it symbolically by pulling back his head slightly.

18. The Head Lower. The neck lowers the head frontally from the neutral position and keeps it in its drooped posture. Like the Head-Aside action this is a way of cutting off the outside world, but because it embodies a general height reduction it has a depressed subordinate air about it. The head swept away to one side can be haughty but the lowered head cannot. It usually looks positively downcast, but on certain occasions it may also be employed to signal that 'I am pensive' or 'I am deep in thought, so do not disturb me'. Modesty and shyness can also be signalled by a sudden lowering of the head to hide the face. In a hostile context, the lowering of the head may take on a totally different meaning, signalling an imminent head-charge. In such a case a key difference is that the eyes stare forward, glaring at the enemy, instead of looking downwards with the rest of the face.

19. The Head Lift. The neck raises the head from a lowered position to the neutral posture. Whereas head lowering could be used as a method of social disengagement the Head Lift acts as a deliberate engagement device. The subordinate who has entered a room and is now standing in front of a dominant individual notices that his head is lowered and that he is writing at his desk. If he is sufficiently cowed by

■ *In many ways the neck is the most 'holdable' part of the body, a favoured region for clasping or grasping in moments of both love and aggression. Couples kissing will frequently clasp the partner's neck as a way of increasing the pressure of the lips and, at the same time, indicating the desire to keep the two bodies in close contact for a longer period. Over-used, such neck-clasping can soon become cloying and oppressive, giving the partner a sense of being trapped.*

the presence of the great man he will simply stand there silently until the figure at the desk lifts his head and looks at him. This simple movement may be sufficient to trigger the subordinate to speak. The dominant individual does not need to say anything – the Head Lift is all that is required.

20. The Head Tilt-back. The neck tips the head backwards from its neutral position and holds it in this tilted posture. This is the nose-in-the-air posture of the snob or the unusually assertive individual. It is the opposite of the forlorn subordinate Head Lower. The emotions of the person with the head tilted back range from smugness and haughtiness to superiority and defiance. This is essentially the posture of 'challenged dominance' rather than calm dominance. The success of the movement depends on the fact that the level of the eyes is slightly raised, giving the illusion of increased height. Short men who are forced to look up at their companions when talking to them often give the impression of being pompous when they are not, because it is natural for them to adopt the tilt-back posture of the head for purely physical reasons.

If the eyes are not looking at the companion but are also raised up – or perhaps shut – then the message is quite different. Heads tilted back in this way belong to people who are in agony or ecstasy, experiencing a massive dose of pain or pleasure. They are suffering from sudden over-stimulation and their response is to cut-off from the world around them. If the sensation is extreme enough, then regardless of the exact nature of the emotion the Head Tilt-back occurs and with it a temporary release from the source of the input.

21. The Head Cock. The neck tilts the head to one side and holds it there. It is done while facing the companion, at a short distance. This action derives from a childhood comfort-contact in which a child rests its head against the body of its parent. When an adult (usually a female) tilts the head to one side it is as if it is being leant against a now-imaginary protector. This 'little child' act is contradicted by the mature sexual signals of the adult body of the gesturer, giving the Head Cock an element of coyness. If used as part of a flirtation the Head Cock has about it an air of pseudo-innocence or coquettishness. The message says 'I am just a child in your hands and would like to rest my head on your shoulder like this'. If used as part of a submissive display the gesture says in effect 'I am like a child in your presence, dependent on you now as I was when I laid my head on my parent's body.' It is only a mild signal, however, and does not make this point strongly but merely hints at it. In this submissive form the Head Cock is popular among old-fashioned shop assistants and obsequious head-waiters who wish to increase the feeling of superiority in their clients.

22. The Head Loll. The neck allows the head to droop downwards and sideways. Performed as a signal, this is a 'mock-falling-asleep' sign indicating intense boredom on the part of the gesturer.

There are many more head movements and postures created by the human neck muscles as specific social signals, but this collection is sufficient to illustrate the great subtlety and complexity of the ever-shifting neck. Anyone who has suffered from a stiff neck or been forced to wear a medical collar following neck injury will know just how deprived the human frame feels when it cannot express itself with this part of the body.

THE SHOULDERS

■ The main function of our shoulders is to provide a strong foundation for our multi-purpose arms. Ever since our ancestors adopted an upright way of life our 'front legs' have become increasingly versatile, and our shoulder girdle, or pectoral girdle, has had to serve that versatility by becoming more flexible. The collar bones and shoulder blades are capable of movements through about 40 degrees and, with their powerful muscles, can help the arms to swing, twist, lift and rotate in an amazing number of ways.

One of the most important early tasks of these mobile arms was weapon-toting, not for war but for hunting. This became a male specialization and it followed that males needed stronger arms than females. It followed again from this that male shoulders had to be more massive, with the result that in the shoulder region we see one of the most striking (non-reproductive) gender differences of the human body. The male shoulders are much broader, thicker and heavier than those of the female, a difference exaggerated by the female's wider hips. The typical male body shape tapers inwards as it descends while the typical female shape broadens out.

Inevitably this sex difference led to a variety of cultural exploitations. If men wished to appear more masculine, they had only to add some kind of artificial width to their shoulders. The most obvious example of this is the military epaulette, which both stiffens the line of the shoulder and adds to its width with projecting 'ends'. To draw attention to this masculine feature, special badges or emblems of rank are often added to the epaulettes, obliging the eye to dwell longer on the enhanced shoulder-shape.

◄ The strong shoulders of the human male are capable of bearing heavy loads, but their primary function is to provide a foundation for all-important arm movements. The increased mobility of the arms and their new duties of throwing and hitting have put extra demands on the male shoulder musculature, so that during evolution this has become sturdier and broader.

The shoulders as accentuated gender signals.

■ *When our early male ancestors became hunters they needed more-powerful shoulders to provide a support-system for their stronger weapon-toting arms and hands. This meant that broad shoulders became a male gender signal and so, by contrast, narrow shoulders became a female signal. The shoulders of an average female are approximately seven-eighths as broad as those of an average male – but breadth is not the only way in which they differ. Even more important is their measurement from front to back. In this direction the difference is even greater, reflecting the comparative weakness of the female shoulder musculature.*

The most exaggerated forms of male shoulder 'expansion' today are found in two very different contexts: Japanese theatre and American football. In the traditional Japanese Kabuki theatre the strong serious male roles are played by actors wearing vast 'wings' of stiff brocade which almost double their real shoulder width. This garment, known as the *kami-shimo*, gives them an immediate air of domination and authority. American footballers, with their massively padded shoulders, also appear threatening and overpoweringly masculine, even when they are standing inactive at the side of the field. Our response to these body proportions is so deep-seated that we simply have no choice about it.

Women who have wished to assert themselves have also adopted artificially broadened shoulders, and this has happened at several points in the recent past. It was noticeable in the dress of the Eman-

◄ As with many gender signals, artificial exaggerations of male shoulder size have been common. This American football player looks intensely masculine because of the width and thickness of his protective shoulder pads. In addition to their size, they also appear to be hunching up in a primate threat posture. Male chimpanzees actually erect the hairs along their shoulders to enlarge their hunched up shoulder-lines when threatening rivals, and it is significant that the sparse hairs which human males still possess in this region grow in an unusual upward direction, suggesting that back in our hairier days we too had a similar display.

■ If shoulder muscles are exaggerated to excess as sometimes happens with the more fanatical bodybuilders, they lose their masculine appeal for many women. This is because the amount of time and effort required to develop them to this pitch implies a degree of narcissism which makes any female feel excluded. It also suggests an obsession with the purely physical.

▲ Military men have often exaggerated their manly proportions by means of shoulder pads and epaulettes – and for some comedians the temptation to caricature this particular male conceit is irresistible.

cipated Woman of the 1890s. In her bid for sexual equality she displayed her mood by adopting 'shoulder equality'. Fashion historians have recorded the shift: 'The slightly puffed shoulders developed into epaulettes and then into something looking like small bags until by 1895 they were rather like a pair of large balloons quivering on the shoulders.' These broad-shouldered women competed with men by taking university degrees, going out to work and engaging in sports that had hitherto been barred to them. Beneath their masculine attire, however, they still wore corsets and petticoats. They were masculine in public but feminine in private.

The second wave of big-shouldered women appeared during the Second World War, when rather square-cut, military-style clothing was adopted even by civilians. This included a stiffly-supported shoulder line which often extended well beyond the natural end of

■ *Because male shoulders are broader than female shoulders, it follows that whenever women wish to assert themselves they can display their mood by adopting fashions which artificially widen this region of the body. Alternatively, individual females who possess unusually broad shoulders can be selected as images symbolizing an assertive female trend.*

the shoulders. It was an appropriate display for a wartime period when women were playing a bigger role in hostilities than ever before.

The third wave has arrived recently with the women's liberation movement, in the form of what could be described as 'terrorist chic'. Pseudo-battledress tops with epaulettes create the required air of female toughness, and once again the shoulders are squared to give an air of masculine strength. An accompanying shift in glamour figures can also be detected. Leading ladies in films no longer mince and wiggle, they stride and jut. Girls with naturally wide shoulders are given chances that would have been denied them in the 1960s or earlier. Furthermore they resent labels such as 'ladies' or 'girls' as being far too feminine. Figures with broad shoulders prefer to be thought of as one of the 'guys'. As an extension of this trend female body-building has surfaced and found a considerable following. A few decades ago a female musclewoman would have been viewed as a circus freak, but in today's climate she has become a symbol of the new female strength, and has the powerfully developed shoulders to prove it.

One aspect of male shoulders which is difficult to imitate is their height. The average male is about five inches taller than the average female, with the result that males have always been able to offer a shoulder to cry on, not because it is broad but because it is high enough to act as a comfortable resting place for a distraught female cheek. With tears and vulnerability out of fashion, the liberated, battledressed female still finds herself faced with a tall shoulder. Since the extra height of her male companions evolved through their primeval hunting activities it seems unfair that modern-day desk-bound, pen-pushing males should still display this badge of physical superiority. Unfortunately evolution works at a very slow pace. Another million years of pen-pushing may correct matters, but in the meantime male shoulders will remain stubbornly at female head-resting height. Short of cutting down men's legs the only hope of shoulder-height equality is the wearing of five-inch lifts on female shoes. The difficulty with this is that very tall shoes make for instability and the need for a helping male hand, which defeats the object of the exercise. For the time being it looks as if, physically, women will have to continue to look up to men, even though mentally thay have adopted a very different viewpoint.

The mobility of the human shoulders is such that even when they are not involved in arm movements they can be raised and lowered, rounded and squared, hunched and shrugged. Some of these movements have become modified as special signals in the language of the

Square, high and hunched shoulders.

body, but to understand them it is necessary to look at the more primitive reasons for adopting one type of shoulder posture rather than another.

Basically, shoulders are kept *down and back* in a mood of calm and alertness and are brought *up and forward* in moments of anxiety, alarm or hostility. The cheerfully resolute, dominant individual keeps his or her shoulders lowered and squared. Individuals who are being dominated or who are frightened or angry tend to hunch up their shoulders as an act of self-defence. If someone is threatening to hit us over the head we automatically try to protect the head and neck region by tucking the head down into the shoulders, and this tensed-up position has become synonymous with unpleasantness of every kind.

It follows from this that if we have a stressful day at work, where we are liable to repeated disappointments or irritations, we will keep on tensing-up our shoulders. These actions might be useful if we were being beaten with a stick, but they are useless when we are being 'beaten' instead with harsh words. At the end of such a day we return home slightly more round-shouldered than we were in the morning when we set out. If this is repeated day after day, week after week, we can eventually develop a marked stoop in our posture, with hunched, permanently rounded shoulders. The long sloping neck we displayed as children slowly shrinks down into our shoulders until it all but disappears. By old age, our chin has come to rest on our chest.

Successful individuals (by which I mean successful to themselves as well as to the outside world) do not undergo this gradual hunching-up decline, and there have been plenty of ram-rod-straight ninety-year-olds to prove the point. Full of self-confidence and optimism, there are too few blows in their lives to make them duck in anticipation into a life-long hunch. For others, and they are in the majority, there are too many anxieties in modern living to avoid at least some degree of shoulder-tensing during waking hours. For them a soothing massage of the back of the neck and of the powerful shoulder muscles can work wonders at the end of a hard day. Despite this, and despite the presence of a loved one who could easily spare a few minutes to ease the tensed muscles, such physical comfort is comparatively rare in ordinary family life. Part of the reason for this is that to receive such comfort is to admit to the existence of a status problem. To ask for relief from aching shoulder muscles is to ask for relief from failures and frustrations we would rather forget. This is a pity because, apart from the aching shoulder muscles, a bad posture in that region can lead to poor respiration and a variety of other complaints. A relaxed dominant shoulder posture literally takes a weight off the body and enables it to function more efficiently.

Two special shoulder signals owe their origin to this defensive hunching. They are the 'Shoulder Shake' and the 'Shoulder Shrug'. The Shoulder Shake is a conspicuous addition to social laughing. If something makes us laugh when we are on our own we let out a guffaw or a snigger, but we do not usually add body movements. These are kept for social occasions, when we are not only amused but also want to convey our amusement to our companions. Because there is a slight hunching up of the shoulders when we laugh we can augment our display of good humour by exaggerating this hunching action,

■ *As we grow older, our predominant shoulder position becomes an irreversibly fixed posture. If we have slumped too long in our easy chairs, or if we have experienced a life too full of moments of anxiety and tension, then our repeated shoulder-hunching may one day become a permanent fixture, bending us over in the sad stoop of old age. Others, with a more mobile, vigorous approach to living and with a more light-hearted attitude to existence, will keep a more flexible frame and a more upright posture, with relaxed shoulders that will see them through a much greater life span.*

The reassuring power of the shoulder embrace.

repeating it and enlarging it, so that the shoulders rise and fall rapidly with our laughter. With some individuals this reaches a level where it almost becomes a personal trademark, as in the case of former British Prime Minister Edward Heath.

The reason why people 'shake with laughter' in this way is that the basis of humour is fear. Humour must shock us in a safe way and we signal our surprise and our relief simultaneously by laughing. The hunching of the shoulders that accompanies this vocalization is part of the primeval fear element. The repeated rise and fall of the shoulders when they shake with laughter is saying, in effect, that there *is* fear present, but it is not serious. If it *were* serious, the shoulders would stay up.

The Shoulder Shrug has a similar origin. In this action, the shoulders are raised into a fully hunched position for a brief moment and then dropped again. The hands are turned palm-up in a manner similar to that seen in begging or imploring, and the mouth corners are lowered. Sometimes the eyes are deflected upwards as if avoiding your gaze. This combination of actions indicates a momentary lowering of status, a symbolic impotence, a fleeting acceptance of an inability to cope.

Most shrugs are signals of ignorance ('I don't know'), indifference ('I couldn't care less'), helplessness ('I can't help it') or resignation ('There's nothing to be done'). They are all negatives, admissions of inability, and with that inability goes the brief loss of status. As the status momentarily sinks, the shoulders momentarily rise. This formal adoption of a tensed-up posture does not mean that we are seriously stressed or that we necessarily feel inferior to, or threatened by, the speaker who has provoked our shrug. It simply means that we cannot deal with his specific question or comment.

The use of shrugging varies considerably from culture to culture but it always has the same basis. In some Mediterranean countries its threshold of use is very low. One only has to make a passing mention of government restrictions, taxation or the failure of the local football team to produce an immediate, prolonged and silent shrug. This expresses the shrugger's complete helplessness in the face of inconceivable folly. His shrugging posture says: 'These blows keep falling on my poor shoulders and I raise them like this to protect myself, but what good does it do?' In countries farther north shrugging, like other gesture-replies, is considered impolite and it occurs less frequently, but when it does appear it has similar roots.

Not all shoulder-raising is defensive hunching. There are several kinds which do not involve a protective element. Raising and rounding the shoulders forward, with the arms clasping the front of the body is a form of 'vacuum embrace'. It is the action of hugging ourselves in the absence of someone else to hug. Here the raised and rounded shoulders are mimicking the posture they would adopt if the loved one were present to be truly hugged. Another version of this is the raising of one shoulder to make contact with the cheek. The head rests against the shoulder, again as if performing a tender action towards a loved one. The shoulder is 'standing in', as it were, for the missing shoulder of the absent loved one.

There are comparatively few regional gestures employing the shoulders. Among the more interesting are the following:

▲ *The action of hugging or embracing the shoulders gives us a powerful sense of security because it enables us to relive briefly, as adults, the infantile sensation of clinging to the large bodies of our parents. When we were tiny children the primeval primate pattern of clinging to the adult body provided the most powerful feeling of protection and loving care known to us. We did it when we were scared, for safety, and when we were happy, to celebrate our mood by sharing it. As adults we use it in the same two ways, and if there is nobody available for us to hug we perform the auto-contact response of self-embracing. When we do this we turn ourselves briefly into two people: the shoulders are ours, but the arms are symbolically those of somebody else who, in our imagination, embraces us to make us feel snug and safe.*

◄ *When the other person is real instead of imaginary we embrace their shoulders greedily, like young chimpanzees leaping on their mother's fur and holding tight. Looking at the hugging couple, it is clear that the usual combination is for one to embrace the shoulders of the other, while the reciprocal action is more of a back embrace. If the couple are about the same height the one employing the shoulder version is usually the initiator of the contact, the back hug being the 'replying action'. Only if the initiator is much the smaller of the two does he or she start with the back embrace, to which the bigger person then replies with the shoulder contact. There are many variations of this type of 'allo-contact'.*

▲ *Because it requires the kisser to lower his head, the shoulder kiss has about it an air of servility, when displayed in public. Performed light-heartedly, it acknowledges the sexuality of an off-the-shoulder costume.*

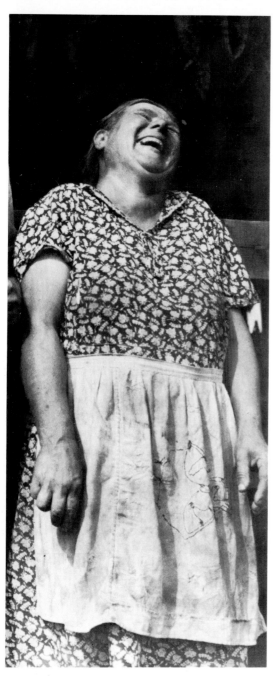

▲ The reason why we hunch up our shoulders when we laugh is that all humour is tinged with fear. What makes us laugh is something that shocks us but which we know at the same time is a 'safe shock'. It will not hurt us and we laugh from relief. But our shoulders still respond to the joke in a primeval way, rising up as if to protect us from some sort of blow or attack.

Why we hunch up our shoulders when we laugh.

Throwing salt over the shoulder. In certain countries it is considered extremely unlucky to spill salt during a meal. To counteract this bad luck it is necessary to throw three pinches of salt over the left shoulder. The reason for involving the shoulder in this superstitious act is not at first clear, but it becomes so once the symbolism of salt is understood. For centuries salt has stood for purity because of its colour, and for immortality because it prevents decay. From ancient times it has been used in ceremonies as a protection against the forces of evil. To spill this precious material is therefore unlucky and will inevitably attract evil forces to the scene. These forces, the devil in particular, never attack bravely from the front, but sneak up from behind, always approaching from the left, which is the unclean, devil's side of the body. So at the moment you spill the salt there is a danger that the devil may be approaching you from behind your left shoulder; and the only way to counteract this advancing menace is to throw purifying salt over your left shoulder, straight in the devil's evil face. Despite the fairytale sound of all this it is surprising how many otherwise worldly people still insist on throwing some spilt salt over their left shoulders after a moment of clumsiness at table.

Patting oneself on the shoulder. This action mimics the congratulatory pat on the shoulder one would expect from a companion. It signifies pride in some achievement and is always done in a jokey way, although it may carry with it an implied mild criticism of one's companions for not having offered their own congratulations.

Brushing one's shoulder. In South America this gesture, in which the person brushes imaginary dust from his shoulder, implies that someone is behaving in a servile, fawning way to curry favour. The gesture mimics the toady's action of fussing around a dominant individual for whom he is even prepared to brush a tiny speck of dust from a shoulder.

Shoulder striking. Among Eskimos a playful blow on a friend's shoulder with the palm of the hand has become a stylized greeting, and an informal version of it can be seen in many other societies. When two friends meet and some body-contact is called for, a mock punch on the shoulder is sometimes used. This is done when a handshake would seem too pompous and an embrace too intimate. It is not common but often surfaces between friends who have had a 'tough' relationship of some kind in the past, such as participation together in military service, sports events, or heavy drinking sessions. Like all mock attacks it signals closeness without softness. It indicates that the friends trust each other enough not to mistake the action for a real act of aggression.

Shoulder arm-cross. Finally there is a gesture seen today only in Hollywood epics about ancient Egypt. It is an act of submission to a dominant individual and is often accompanied by bowing. In it the subordinate individual makes contact only with his own body. He greets the powerful one by placing his left hand on his own right shoulder and his right hand on his own left shoulder, the arms crossing in front of his neck. This appears to be a mimed act of 'vacuum embracing' in which the gesture of the submissive individual says to the dominant figure: 'I embrace you like this but I am not worthy to approach close you.' The subordinate's hands rest respectfully on his own shoulders as substitutes for the shoulders of the great one.

THE ARMS

■ To any four-footed creature the human arms must look like a pair of useless front legs dangling in the air. But when our ancestors stood up on their hind legs our forelimbs were dramatically relieved of load-bearing and able to specialize as multipurpose manipulators. Our front feet turned into sophisticated prehensile grippers and our front legs became their wonderfully mobile servants.

The arms serve in two ways: with power and with precision. If the hands must act forcefully – climbing, throwing, beating, punching – the strong arm muscles, such as the biceps and triceps, ripple and bulge into action. If the thumb and fingers are acting with delicate precision the arm then operates as a mobile crane, moving the hand into the ideal position for the more finicky work to be carried out.

The arm is based on three long-bones: the heavy *humerus* of the upper arm and the lighter *radius* and *ulna* of the forearm. These bones are apparent at the shoulder, the elbow and the wrist, but elsewhere they are embedded in muscles. The two bones of the forearm cross over each other when the hand is rotated into a palm-up position, which means that the most relaxed arm position is a palm-down one. For those who cannot remember which bone is the ulna and which the radius, the ulna is the slightly more slender one in line with the little finger while the radius is the stouter one in line with the thumb.

The main arm-muscles and the movements they create are as follows: The *deltoid* is the bulky muscle that rounds off the top of the arm where it meets the shoulder. Its function is to raise the arm up and away from the side of the body. The *biceps* are the bulging muscles on the front of the upper arm. Their function is to bend the arm. The *brachialis* is a smaller muscle at the lower end of the biceps and has a similar function. The *coraco-brachialis* is another of the smaller muscles, this time at the top end of the upper arm. It rotates the arm inwards and helps to bring it to the side. The *triceps* are the powerful muscles at the back of the upper arm, with the function of extending the forearm. The *flexor* and *extensor* muscles of the forearm bend and stretch the fingers and hand; and the *pronator* and *supinator* muscles rotate the forearm into palm-down and palm-up positions respectively.

Muscle-building techniques make it possible to pump these arm muscles up to an amazing degree and the bulging arms of competitors in the 'Mr Universe' and other such contests create an impression of immense masculine strength. Despite this, many females report that they do not find such displays sexually attractive. The main reason appears to be that the amount of effort obviously involved in developing the arms to this extent implies a degree of self-obsession verging on the narcissistic. The top body-builder seems to be less interested in the body of a female companion than in the body he sees in the mirror.

Although excessively developed arms may fail to excite the female eye, the more moderately proportioned 'strong-arm' male does transmit important gender signals. The normal sturdy male arm is a powerful masculine signal which contrasts strikingly with the slender upper limbs of the typical female. Female arms are both shorter and weaker, and their forearms are proportionally shorter. The longer forearms of the male are seen as reflections of his specialized evolutionary role as an aimer and thrower. As a result of this, men are much better javelin throwers than women. The male world record for this

■ *During the course of evolution, human arms developed much greater strength in the male than the female. This was due to the new division of labour between the sexes, with males being involved in hunting activities that required vigorous throwing and the delivery of powerful blows. As a consequence of this, modern sporting events show their greatest gender differences in achievement when arm-strength is demanded of the contestants.*

▲ Another gender difference in the arm region is the way in which the elbow bends.

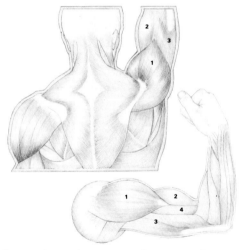

▲ Among the arm's more conspicuous muscles are the deltoid (1), the biceps (2), the triceps (3), the brachialis (4) and the various superficial extensors of the forearm.

▶ The hairy human armpit has suffered a great deal in recent years, shaved, sprayed and scented to prevent it from transmitting its biological odours. These odours, pleasant enough in the freshly washed adult, become unpleasant when they grow stale, which easily happens now that we wear clothes and create an unnatural 'hothouse' atmosphere in the armpit zone. Fresh and unstaled, the armpit secretions are strong sexual stimuli, and certain old folk-customs reflect this. At village dances, young men would keep a handkerchief tucked into a sweaty armpit and then flourish it under the noses of the girls they wished to seduce, a practice which may explain the waved handkerchiefs in some forms of revived folk-dancing.

Even the elbow shows a sex difference.

event is 96·72 metres (317 feet), the female 72·40 metres (237 feet). This difference is much greater (33 per cent) than the differences for track events (average 10 per cent).

Another gender difference concerns the elbow joint. In females the upper arms are naturally closer to the flanks than is the case for males. The broader shoulders of the males mean that their arms hang down away from the body. When they are allowed to dangle in space they have a strongly masculine flavour, but if a male were to force them close in to his sides while spreading his forearms out away from his body he would appear effeminate. This is because there is a greater elbow angle – greater by about 6 degrees – in the female. So the posture of the arms also provides us with significant gender signals that cannot be ascribed to local cultural conditioning.

If the elbow suddenly encounters a hard object, there may be a numbing, stinging sensation accompanied by considerable temporary pain. This is described as 'hitting one's funny bone', a phrase which is a play on the technical term for the bone involved, the *humerus* of the upper arm. It is the knob at the lower end of this bone which is the 'funny bone', because it is there that the ulnar nerve is exposed just under the skin. It is striking this nerve that causes the stinging sensation and for a moment incapacitates the arm in question.

The word elbow comes from the combination of ell and bend. An ell was an ancient measurement equal to two cubits and a cubit was the length of the forearm, from arm-joint (elbow) to fingertip. Other early measurements were based on the length of the human arm, and the yard was said to have been established as being the length of the arm of Henry I, the twelfth-century king of England. A rival claim has the yard as the distance between the nose-tip and the tip of the outstretched middle finger of Edgar, the tenth-century king of England.

Another anatomical detail of the arm which deserves brief mention is the much-maligned, much-shaved and much-sprayed armpit. Technically known as the axilla, this small hairy zone plays an important role in chemical signalling and reflects a major change in the sexual habits of the human species. When our remote ancestors mated, with the female on all fours, the armpits were nowhere near the partner's face. When eventually we adopted a vertical gait and switched to a face-to-face posture as the dominant sexual position, embracing couples found their noses close to the shoulder regions of their partners. Nearby was the partly enclosed armpit, the ideal site for the development of specialized scent glands.

The armpit scent glands are called apocrine glands and their secretions are slightly oilier than ordinary sweat. They do not develop until puberty, when the arrival of sex hormones activates them and at the same time causes the growth of armpit hair. The hair acts as a scent trap, keeping the glandular secretions within the axillary region and helping to intensify their signal.

There is an old English folk-custom which says that if a young man wishes to seduce a girl at a dance he must place a clean handkerchief in his armpit, beneath his shirt, before starting to dance. Afterwards he takes it out and pretends to fan himself with it in an attempt to cool off. In fact what he is doing is wafting his apocrine scent over to the girl, who is promptly seduced by its fragrance. Considering the major industry which today is based on the sale of underarm deodorants,

Fresh armpit secretions are sexual stimuli.

this story sounds rather odd. If the human species carries such a valuable sexual stimulus under the arms, why do so many go to such trouble to remove it, with washings, rubbings, sprayings and, in the case of fastidious females, depilations? The answer has to do with clothing. The young man in the folk-tale, well scrubbed and wearing his best clean shirt for the dance, produces *fresh* apocrine secretions from his scent glands. Soaked in these his clean handkerchief really does carry a strongly sexual scent signal. That is the primeval system at work. Sadly, today, with our bodies covered in layers of clothes, our sweaty skin can easily become a hot-house for the breeding of millions of bacteria. Our natural body fragrance turns sour in this unnaturally confined environment and our scents become stinks. The unpleasantness when this happens concerns us so much that we would rather spray our armpit glands into abject submission than risk our axillary attractions turning into the socially dreaded 'body odour'.

Recent researches seem to indicate that the armpit secretions of males and females differ chemically and presumably have odour appeal specifically directed at the opposite sex. The male's is said to be muskier, but in its pure fresh form neither secretion is easily detected *consciously* by the human nose. They appear to act at an unconscious level, leaving us feeling stimulated but not knowing quite why. Orientals, incidentally, lack this underarm odour-signalling system almost entirely. The Koreans are the most extreme group in this respect, at least half the population having no axillary scent glands at all. The glands are also rare among Japanese – so much so that to have strong-smelling armpits in Japan is considered a disease and has been given the technical name of 'osmidrosis axillae'. At one time men suffering from this 'ailment' were even excused military duty.

Turning to arm postures, there are four main ones to consider: Arms Down, Arms Up, Arms Spread and Arms Forward. (Cases where an arm movement brings the hand into contact with some other part of the body such as the nose or the mouth are dealt with under those headings.) The Arms Down posture is the neutral one, with the arm muscles at their most relaxed and inactive. As part of the balancing act of bipedal locomotion we swing our arms slightly out of this resting position as we stride along but, unless we have been bullied into a showy military marching gait, we do not put much effort into this action. Even after a long cross-country walk, when our feet are aching and our leg muscles are exhausted, our gently moving arms still feel fresh and relaxed. It is only when we start to pull them right away from the body that they feel the strain of our endeavours.

The Arms Up posture is the hardest to hold for any length of time. It is the typical gesture of triumph and victory much loved by politicians and footballers. With their arms stretched aloft they greet their followers and celebrate their high status with a 'high' posture. The raising of the arms makes them seem taller and stronger and also renders them more conspicuous at moments when they most want to be seen. They only hold this position for a matter of seconds, however. If they tried it for a matter of hours or even minutes instead of seconds, they would soon find themselves suffering. This was what Moses experienced when he stood on a hilltop watching a battle between the Amalekites and the children of Israel. As long as he kept his arms up high, symbolizing victory, the Israelites did well; but as soon

The overarm blow is the primeval attack movement of the human species.

■ *Contrasting with the grab-and-bite action of the apes, the overarm blow is the primeval attack movement of the human species. It also relates to our ancient hunting techniques, which involved overarm throwing of weapons. We often recreate such actions in modern sports. In fact, most of our present-day sports are based on aiming and running – the basic ingredients of the ancestral hunt.*

as his strength began to flag and his arms sank down the Amalekites began to gain the advantage. Aaron and Hur solved the problem by standing on either side of him and holding his arms aloft until sunset, when the Amalekites were finally defeated. The posture of raised arms had a 'magical' power to lift the spirits of those who saw it.

A totally different meaning applies when a man with a gun commands 'Hands up!'. Here too the arms are raised, but in defeat rather than victory. There is a subtle difference, however, in the angle of the arms in the two cases. In the victorious posture, the arms are usually kept straight or, if they are bent, are angled slightly forwards. In the gun-threatened posture, the arms are typically bent slightly at the elbow and are in the vertical plane rather than angled forward. The essence of the defeated posture is that it should indicate rather limp, ineffectual arms and hands, removed as far as possible from the body, where a weapon of some kind might be hidden.

The Arms Spread posture is the long-distance embrace-invitation gesture. Someone greeting an old friend who is still some paces away will fling wide the arms and hold them there until they can be wrapped around the friend's shoulders in an emotional hug. This same posture is seen after a circus performer has completed a difficult trick. He or she will fling wide the arms and the audience will immediately respond by applauding. The performer has invited an embrace and the audience has obliged with the only sort of gesture they can manage from their seats in the auditorium. Their action of clapping the hands together is in origin a highly modified form of 'vacuum embrace', in which the feel of the hug has been converted into the *sound* of a symbolic hug.

The Arms Forward position is more complicated. This can either signal rejection, if the palms are pushing forward; or aggression, if the fists are clenched; or begging, if the palms are held face-up. Like the spread arms it can also be an embrace-invitation gesture, and it can transmit a whole variety of other signals, according to the way the hands are being used.

Specialized arm signals include the different forms of waving, beckoning and saluting, each with its own particular flavour. When a leader gestures from a balcony his arm movements can be seen from

The arms are invaluable as long-distance signalling-flags.

■ We raise our arms to make ourselves look taller when we are triumphant or receiving the acclamation of the crowd. We raise our arms to reach up to heaven when we are praying (a much older action than placing the palms together); and we reach out with our arms when we want to embrace the crowd in front of us. Arm movements of these kinds are often global in distribution and appear to go beyond cultural learning.

a great distance. Their exact shape and style indicates something of his mood. The Royal Family Wave is the modestly elevated, rather limp gesture of passive power. The Clenched-Fist Salute of the Communist leader is the fierce sign of active revolutionary power. The Roman Hail and the Nazi Heil are the stiff flat-handed actions of rigid loyalty. The Military Salute – bent elbow, and hand to headgear – is the stylized intention movement of raising the vizor or removing the helmet, an appeasing act meant to cancel the hostile signal otherwise transmitted by the gesturer's appearance. The Papal Wave is a paternalistic, symbolic sweeping up into his arms of the hordes of God's children who gaze up at him. And so on. We use our arms for gesturing when we need a long-distance signal of a cruder kind than those transmitted by our fingers and our facial expressions. In this role our arms act as invaluable body-flags.

In person-to-person contact the arm region is often a focus of friendly non-sexual actions. If we help an elderly stranger across the street we take their arm to support them. If we guide someone through a door we gently direct them by their elbow. If we wish to attract a stranger's attention we tap them on the arm. If in any of these contexts we touched waist, chest, or head our actions would immediately come under suspicion. The arms are the most neutral of body parts in this respect, as free of special intimate significance as any body zone can be. Friends may link arms when walking together, regardless of gender; but if any other kind of contact was made while walking it would promptly signal intimacy of a special order.

Arms have often been tattooed; but the most common form of adornment has always been the bracelet. Bracelets have nearly always been worn by females and it has been suggested that the custom originated as a way of exaggerating the gender signal of the slender feminine arm, the fine bracelets emphasizing the thinness of the arm diameter inside them. A rival suggestion is that they appeal to males because they are symbolic manacles, suggesting the enslavement of the women by their men.

THE HANDS

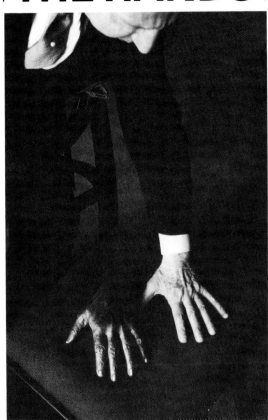

The secret of success for the human hands was the development of opposable thumbs. Freed from the task of locomotion, both on the ground and in the trees, the design of the hands could for the first time become solely manipulatory. This was one of the most important steps in the whole evolution of our species. The human being became dextrous – and crossed a major threshold into a world where nothing was safe from his grasping fingers.

In a physical sense males are much more grasping than females. The average male hand has about twice the gripping power of the average female hand. This is one of the more pronounced gender gaps and reflects the great importance of strong hands to the primeval hunter. The typical male can exert a grip of about 90 pounds, and with special training can increase this to 120 pounds or more. A vice-like grip was particularly useful for fashioning weapons and other early implements, for throwing objects with force and for other activities involving such actions as hammering, twisting, tearing, clinging and carrying. Even today, tasks which benefit from large powerful hands are still male dominated. There are very few female carpenters.

The Power Grip is only one half of the manual success story, however. The other half is the equally important Precision Grip. Power is achieved by opposing the whole of the thumb against the whole of the fingers. Precision is achieved by opposing only the tips of the digits. In this action the female is superior to the male. The big beefy male hands, although capable of great precision when compared with the short-thumbed hands of other species, cannot compete with the delicately agile, small-boned hands of the human female when it comes to finicky tasks. Females have always excelled at sewing, knitting, weaving and the finer forms of decorative work. Before the pottery wheel was introduced women also dominated the important ancient art of ceramics, where nimble agile fingers were so important in shaping and decorating the vessels. Because pottery was the major art form of the prehistoric period it follows that during that long phase in the human story it was the females and not the males who were the important creative artists – a fact usually overlooked by archaeologists and art historians.

This difference in manual precision is not just a matter of women having lighter, thinner fingers. The female finger-joints are also more flexible, a feature thought to be influenced by hormonal factors. It has been argued that this was a special adaptation to the primeval feminine specialization of food-gathering as opposed to hunting. Food-collecting, involving harvesting roots, picking seeds, nuts and berries, and selecting fruits, required the deft quick fingers of the fine-boned, loose-jointed female hand rather than the power-paws of the muscular male. These are not sexist statements; they simply record the physical division of labour which occurred during our evolution and which made men and women slightly less like one another – each better in certain ways. And of course the process of specialization never went too far. Women's hands remained reasonably strong and men's hands were capable of quite fine work. The strongest of women could always tear apart a piece of meat or open a stubborn bottle-top better than the weakest men in any group; and sailors at sea have always proved reasonably adept with a needle and

■ Of all parts of the human body the hands are perhaps the most active, yet we seldom hear of anyone complaining of 'tired hands'. It has been estimated that the fingers bend and stretch at least 25 million times during the average lifetime.

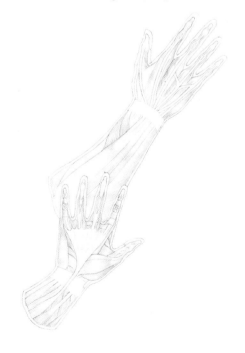

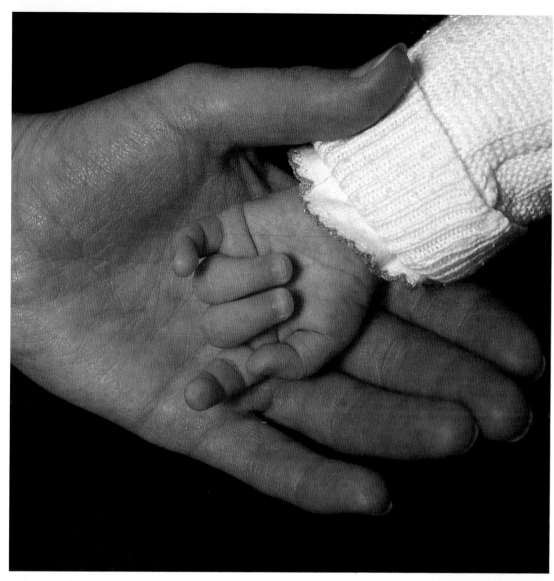

■ With its opposable thumb and its freedom from the demands of locomotion, the human hand evolved into an organ that has literally changed the face of the Earth. Development starts early, as this five-month-old embryo indicates, and during the first year of life outside the womb the baby is already busily fingering the world around it, despite the small size of its hands and the fact that ossification of the bones is not yet complete.

thread. There are even a few male harpists, with remarkably flexible fingers. But right from the earliest days of the Old Stone Age there has been a significant hand-bias – power for men and precision for women.

Of all parts of the human body the hands are perhaps the most active, yet we seldom hear of anyone complaining of 'tired hands'. As complex pieces of machinery they are superb. It has been estimated that during our lifetime the fingers bend and stretch at least 25 million times. Even newborn babies have remarkable strength in their fingers, and their hands are hardly ever still. As they lie in their cots their tiny fingers are flexing and twitching as if anticipating the pleasures of handlings to come. And later in life what handlings they prove to be: typing at a hundred words a minute, playing concertos at breakneck speed, operating complex machinery, performing brain surgery, painting masterpieces, reading braille with the fingertips, and even reciting poetry in sign language for the deaf. Compared with the Rolls Royce of the human hand, the other species do not even own a bicycle.

A pair of human hands contain no fewer than 54 bones. In each hand there are 14 digital bones, 5 palmar ones and 8 in the wrist. The sensitivity of the hand to heat, pain and touch is fine-tuned and there are literally thousands of nerve-endings per square inch. The muscular strength of the hands and fingers comes not only from the musculature in the hand itself but also from more remote muscles in the forearm.

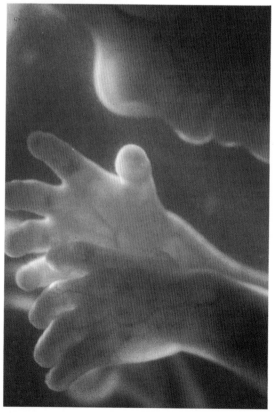

'The hand is the cutting-edge of the mind.'

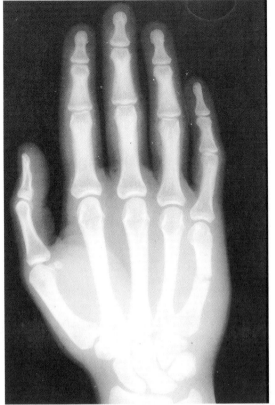

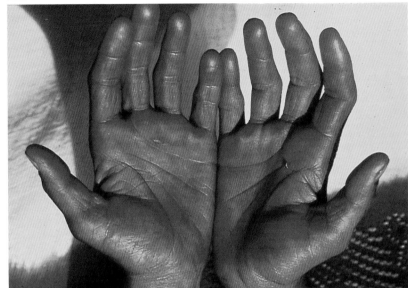

■ The human hand is a brilliant piece of engineering, so complex that no robot imitation has ever come close to copying all its intricate actions. In the eighteenth century Kant called tha hand 'the visible part of the brain' and in the twentieth century Bronowski used a similar image, saying: 'The hand is the cutting edge of the mind.'

◄ Compared with that of the male, the female hand remains closer to the infantile condition. It is more sensitive and less powerful. It is also smaller, the average female hand being almost three-quarters of an inch shorter and half an inch narrower than the male's.

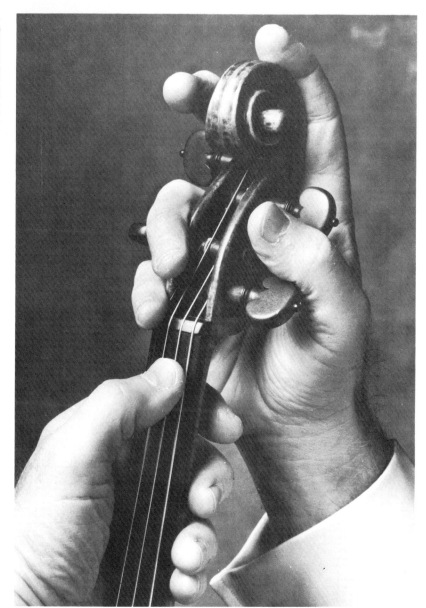

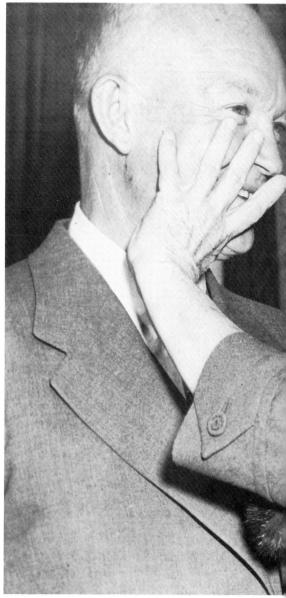

■ The dexterity of the human hand can be seen at its peak in the virtuoso performances of great pianists and violinists. The holding of a violin bow is a supreme example of the Precision Grip, in which the tips of the thumb and fingers are opposed. It was the perfection of this type of grip, as an addition to the primeval Power Grip, which was the secret of the manual success of our species. There are also two minor forms of holding, the Hook Grip and the Scissors Grip. The Hook Grip is used, for example, when we carry a suitcase, employing bent fingers only. The scissors grip is typical of a cigarette smoker, who clamps his cigarette between his first and second fingers without bending them.

On the surface of the hands there are three kinds of lines – the Flexure Lines, the Tension Lines and the Papillary Ridges. The first of these, the Flexure lines or 'skin hinges', are creases which reflect the movements of the hand. They vary slightly from individual to individual, a fact which has ensured a steady income for palmists down the centuries. Like those other confidence tricks, phrenology and astrology, palmistry has lost ground rapidly in the twentieth century and is now at last no more than the fairground fun it deserves to be. Its one useful legacy is the naming of the various crease lines in a way that can easily be remembered. The four main lines are the 'head line' and the 'heart line', running across the palm, and the 'life line' and the 'fate line', running around the base of the thumb. In apes, the head line and the heart line are one, but in humans the independence of the forefinger is such that it splits the line into two. Some people, however, still display the ancient condition – the single line or 'Simian Crease', right across the palm. It is present in roughly one person in twenty-five.

The Tension Lines are the small wrinkles which increase with age and become permanent as the skin loses its elasticity. The tiny Papillary Ridges are the 'grip' lines which give us our finger prints. Sweat makes these little ridges swell and become more elevated, helping us to grip objects firmly.

The sweating action of the hands is unusual. When we are asleep the palmar sweat glands cease activity, no matter how hot we may be in

▲ *As a result of an illness in infancy, Helen Keller, seen here 'reading' the face of ex-president Eisenhower, became a life-long blind deaf-mute. Despite her disability, she fought a heroic battle to communicate with her fingers, and won. She wrote eleven books and travelled the world giving lectures, trying to remove the stigma attached to blindness by society. Her hands became so sensitive that she could even 'listen' to music with them. She did this by placing her fingertips on the membrane of a radio loudspeaker and, in her own words, 'could distinguish accurately between the cornet and the roll of the drum, the deep tone of the cellos and the singing of the violins.'*

■ *At the other extreme from Helen Keller's sensitive fingertips is the massive fist of world champion Mohammed Ali. One-and-a-half times the size of the average male fist, his hands became the most violent weapon in the history of unarmed combat.*

bed. In fact, they do not respond at all to heat increases like the sweat glands on other parts of the body. They only react to increased stress. If your palms are bone dry you are relaxed. As you become more and more anxious they get damper and damper, preparing themselves for the physical action which your system is anticipating. Unfortunately the human body evolved this reaction at a time when most stress *was* of a physical nature, but today our modern tensions are more likely to be psychological ones, with the result that we are left with damp sticky palms and nothing to grip. Palmar sweating is a relic of our ancient hunting past which most urbanites could do without.

During the famous Cuban missile crisis, when the Western World held its breath, fearing a nuclear war, all laboratory experiments into palmar sweating had to be abandoned temporarily. The general increase in stress raised the sweating rate to an extent that made it impossible to get a 'relaxed' reading from any of the subjects being tested. Such is the sensitivity of the human hand.

The fingerprints form three basic patterns: loops, which are very common; whorls, which are moderately common; and arches, which are rather rare. No two human digits have ever been found with identical fingerprint details. Despite popular opinion to the contrary, even identical twins have different fingerprints. The use of prints for identifying individuals is centuries old. Over 2,200 years ago early Chinese businessmen were labelling their seals of authority with their

personal finger marks. Since signatures can so easily be forged, it is surprising that we do not follow the ancient Chinese custom. The use of fingerprinting in modern crime detection has become highly sophisticated, with the technique of 'ridge-counting' and attention to tiny line details bearing such names as 'lakes', 'islands', 'spurs', 'crossovers' and 'bifurcations'. There is no way a criminal can avoid this type of identification by trying to alter his prints. Even if he has them painfully worn away, they soon grow back again, and they do not alter with age.

Racial differences in fingerprints are slight. Caucasians have fewer whorls than Orientals, for example, and more loops; but the differences are only statistical.

There are three special colour qualities of the human hand that have aroused interest. When white-skinned people become suntanned the backs of their hands become brown, but their palms refuse to darken. This special feature of the human palms is said to have evolved in connexion with the need to keep hand gestures highly conspicuous. Even dark-skinned races have pale palms.

Anyone who has been snow-balling will have discovered that after a period of time has passed the palms become bright red. This special response appears to be a mechanism to prevent damage to the sensitive palmar skin from chilling. The reaction to prolonged cold produces a dramatic increase in blood flow which warms the hands. This is a remarkable and complex response. The initial reaction of the hands to the cold snow is vaso-constriction, reducing the flow of blood to the surface. This is the usual reaction of the body as a whole. It prevents warm blood from dissipating the vital body heat from the surface. This response remains the same for the rest of the body, no matter how long the exposure to cold may last, but the hands operate independently in a special way. After about five minutes they switch from strong vaso-constriction to the exact opposite – strong vaso-dilation. The palm and finger blood vessels suddenly expand and the hand goes bright red. Then, after about another five minutes the process reverses itself. If the ungloved snowballer were sufficiently stoical to keep going for an hour he would observe his hands turning from blue to red and back again every five minutes. This is an emergency protection system which we probably evolved back in the Ice Age, when frostbitten hands could easily spell disaster. By repeatedly warming the hand surfaces for brief five-minute spells it prevents the prolonged chilling that can cause the real damage. By repeatedly allowing them to cool for five-minute intervals it conserves precious body heat.

One of the most extraordinary items of human hand-lore is the claim by certain holy people to be afflicted by *stigmata*. These are supposed to be spontaneously-formed wounds on the palms of the hands, similar to those which Christ is assumed to have suffered on the cross. The vast majority of the 330 people on record who have displayed the stigmata have been Roman Catholic clergy – priests, monks or nuns – and the Church authorities have always been uneasy about such claims. It is not the wounds themselves that are in doubt, but whether they were miraculously caused. In typical cases the wounds on the palms would suddenly start to bleed, then heal up, then bleed again. One stigmata-owner bled to a very tight schedule:

■ *The postures of the hands when gesticulating reflect the mood of the persons concerned. Palm-up is a pleading posture, Palm-in an embracing posture, Palm-down a calming posture, and so on. Here, one type is illustrated in a number of different contexts, but with the same basic message – keep back, or keep away. The Palm-front posture is a universal signal of rejection, since it mimes the act of pushing someone back.*

The Palm-front posture mimes the act of pushing someone back.

between one and two o'clock every Friday afternoon, and then again between four and five o'clock. The first recorded instance is that of St Francis of Assisi, who had two Popes as witnesses, and the last was that of the Italian priest Padre Pio, who died in 1968 after collecting £7 million from impressed followers, to build a hospital.

The most likely explanation of the cause of the stigmata wounds, assuming there was no deliberate self-mutilation, is that they were cases of localized virus infection. Children using public swimming pools often catch verrucas – small virus-warts, which have to be removed surgically. Similar warts may also appear on the palms of the hands, although they are less common there. When they do develop, however, they are often scratched and start to bleed. The scratcher may not even recall touching them. After a while they heal, but the process is much slower than in an ordinary cut. Because of the virus,

Loop-patterned fingertips are the most sensitive.

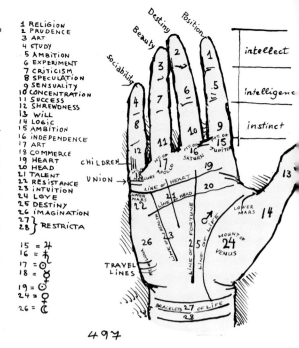

the healing is not perfect and sooner or later they start to bleed again, growing larger in the process. Surgery is needed to remove them permanently. It is easy to see how such a minor affliction could fire the imagination of a devout Christian and become a miraculous re-enactment of Christ's suffering.

Turning now to the digits – each has its own special names, symbolism and gestures. They are as follows:

The first digit: the Thumb. This is without question the most important of the five digits, since it endows the hand with its grip. Its vital role has been recognized since medieval times, when compensation for the loss of a thumb was placed at more than four times that of the little finger. If a thumb is lost today, modern surgery can adjust the forefinger so that it works in opposition to the other fingers, to some extent replacing the hand's gripping action.

In Latin the thumb was known as *Pollex*. In ancient times it was dedicated to Venus, presumably because of its phallic significance. In Islam it was dedicated to Mahomet.

The thumb has three key gesture meanings: it points direction, it delivers a phallic insult and it signifies that all is well. The use of the thumb as a pointer is less common than the use of the forefinger and in ordinary contexts is considered rather surly. An exception to this rule can be seen at the road-side. In modern times it has become the sign of the hitch-hiker, indicating the direction in which he or she wishes to travel. In ancient times it was the downward pointing thumb, aimed at the fallen body of the defeated gladiator, that sealed his fate in the Roman amphitheatre. By jerking the thumb down towards the victim the crowd could indicate their desire to see him slain. If, on the other hand, they wanted him spared they covered up their thumbs. This cover-up later became mistranslated as *turned* up their thumbs, giving us the 'OK' 'Thumbs-Up' sign of recent years. In countries where this mistake did not occur the more ancient meaning of the upward jerked thumb, which is 'Sit on this!', survives to the present day and causes considerable confusion to travellers and tourists. Hitch-hikers in certain Mediterranean countries have been astonished at the anger caused when they tried to thumb a lift from local drivers, little realizing that they were giving them an obscene insult.

Countries where the obscene, phallic thumb gesture is still common include Sardinia, Greece, Lebanon, Syria and Saudi Arabia. It is also found in Australia, where it shares the honours with the OK thumb gesture. Having two thumbs-up gestures in Australia can occasionally cause confusion, but this is usually avoided by stressing the phallic upward jerking movements of the obscene version.

The second digit: the Forefinger. This is the most independent and important of the four fingers. It is the one most used in opposition to the thumb for delicate precision actions. It is the finger that pulls the trigger, the finger that points the way, the finger that dials the phone, the finger that beckons, the finger that calls for attention, the finger that jabs an opponent in the ribs and the finger that presses the button.

It has been given many names. Because it indicates the way, it has been termed the index finger, the indicative finger, the demonstrative finger or the pointer. Because it fires a gun, it has been named the scite finger or shooting finger. It has also at various times been called

■ *There are three basic fingerprint patterns: the arch, the loop and the whorl. The loops are the most common and the arches the least common. Recent studies have revealed that those people showing a predominance of loop patterns have the greatest tactile sensitivity. The prints themselves are the positive impressions left by our papillary ridges when we touch a smooth or soft surface, and no two fingerprints have ever been found which are completely identical. Even so-called identical twins are not identical. And criminals who try to alter their prints never succeed, because they soon grow back again to their original pattern.*

◄ *Palmistry – the telling of fortunes from the crease-lines of the hand – was once taken very seriously, but today has become something of a joke. Because of the fancifully detailed predictions made by palm readers, the whole subject was condemned by the scientific world as a case of fair-ground charlatanism. Recent research indicates that this total dismissal may have been too sweeping, because it now emerges that there are, after all, links between certain hand patterns and some bodily characteristics. Sufferers from Downs Syndrome, for instance, have an extra crease along the base of the palm, and cancer-prone individuals also tend to show certain special types of palm-prints.*

the Napoleonic finger, the finger of ambition, the towcher (toucher), the foreman, and the world finger. Its strangest title is that of the poison finger. In early days it was forbidden to use this finger for any sort of medication because it was believed to be venomous. This probably stems from the use of this digit in aggressive pointing and finger-jabbing, which gives it the symbolic role of a dagger or sword – something dangerous which may wound you, like the stabbing fangs of a snake.

Catholics dedicate the forefinger to the Holy Ghost, Islam to the Lady Fatima. In chiromancy it was given to Jove, in palmistry to Jupiter.

Despite its importance it is usually only the third-longest of the four fingers, being exceeded in most cases by both the middle finger and the ring finger. In 22 per cent of males and 45 per cent of females, however, it is the second-longest finger, relegating the ring finger to third place. Why there should be a significant gender difference in this respect is a mystery.

Apart from the gesturing roles mentioned above, the forefinger is also used in several obscene contexts. The best known of these is the 'Pistola' in which it is pushed through the fingers of the other hand or through a ring made from the curled thumb and the other forefinger, like a penis being inserted in a vagina. This is such an obvious mime that it is understood almost anywhere in the world, and in one instance at least has led to the death of the gesturer. The case is a curious one because it involves the creation of the only obscene banknote in history. When the Japanese invaded China just before World War II they set up puppet banks in certain Chinese cities. Although these were controlled by the Japanese, Chinese engravers were engaged to make the new bank notes. One of these engravers was so outraged by his task that he added a small detail which at first went unnoticed. The elderly sage depicted on the note, whose hands should have been in a formal posture of reverence, was instead shown making the obscene forefinger gesture. The Japanese authorities eventually tracked down the rebellious engraver and he was publicly decapitated, a high price to pay for the satisfaction of making a rude gesture.

Among Arabs there is another obscene forefinger gesture which may also bring swift retribution if used unwisely. It looks innocent enough, consisting of no more than the tapping of a forefinger against the bunched tips of the digits of the other hand. In this instance the forefinger is not being used as a phallic symbol but as a symbol for the mother of the person at whom the gesture is aimed. The five digits of the other hand symbolize males with whom the mother has copulated, and the verbal message of the gesture is 'You have five fathers!' There is no record of it ever having led to the death of the gesturer, but it seems likely.

Other forefinger gestures are less provocative. The holding erect of this digit is often observed and although it has several meanings none is obscene. Apart from calling for attention it can also signal 'I want just one', 'I am Number One', or 'There is one God'. The second of these has become popular among sportsmen recently, when they wish to celebrate the fact that they are for the moment 'Numero Uno'. The third has been favoured by certain groups of born-again Christ-

Like the soles of the feet, the palms of the hand refuse to become suntanned and they remain pale-coloured even in the darkest-skinned races. This has the effect of making hand movements more conspicuous during gesticulation.

ians, where people felt the need for some new gestural symbol.

The third digit: The Middle Finger. This, the longest of the digits, acquired a whole range of names in ancient times, being known variously as the Medius, the Famosus, the Impudicus, the Infamis and the Obscenus. The reason for most of these names is that it was this finger which was employed in the most famous of Roman rude gestures. In this the other digits are bent tight while the stiffly erect middle finger is jerked vigorously upwards. The two bent digits on either side symbolize the testicles and the middle finger is the active phallus. This gesture has survived well during the two thousand years since it was made in the streets of ancient Rome and in modern America is simply known as 'The Finger', although in the past few years it was re-christened 'The Rockefeller Gesture', following its rather conspicuous use on television by someone of that name. An early author, writing several centuries ago, gave an additional reason for this finger being loaded with sexual significance, when he wrote: 'With it are touched the privy parts...'

In the more rarefied atmosphere of religion and superstition, the middle finger has quite different associations. In Catholicism it is the finger dedicated to Christ and salvation; in Islam it is Ali, husband of Fatima; and in chiromancy and palmistry it is given to Saturn, the Roman god of sowing.

The fourth digit: The Ring Finger. This finger has been used in healing ceremonies for over two thousand years. In the ancient rituals of the Aegean it was encased in a finger-stall of magnetic iron and used in 'magic medicine'. Later this idea was adopted by the Romans, who called it the *digitus medicus* – the medical digit. They believed that there was a nerve running from the ring finger straight to the heart.

They always used this finger when stirring mixtures because they thought that nothing poisonous could touch it without giving due warning to the heart. This superstition lasted for centuries, although the nerve to the heart sometimes became a vein and sometimes an artery. In medieval times the apothecaries were still religiously using this finger to stir their potions and insisted that all ointments should be rubbed on with it. The forefinger was to be avoided at all costs. For some, simply to stroke the ring finger over a wound was enough to

■ We usually think of thumb gestures as simple opposites – up for good, and down for bad, but it is not always that simple. In some countries the upward-jerked thumb is an obscenity meaning 'sit on this', which can be confused with the ordinary 'thumbs-up' sign meaning 'OK'. Also the thumb up and down from the ancient Roman games has been misunderstood. The original signals *were* thumb out *for kill* – miming the action of thrusting in the weapon – and thumb cover-up *for* 'spare the gladiator'. This has become mistranslated as thumb down and thumb up.

■ The forefinger is the busiest digit in the language of gestures. It has become the human 'pointer', used in a wide variety of contexts to indicate direction. It is also employed as a small, symbolic club by aggressive or pompous speakers who 'beat' their opponents over the head, when they wag a stiff forefinger at them during angry exchanges. For sportsmen, this digit, being the first finger, has come to symbolize the idea of being 'No. 1', or 'Numero Uno', and it is becoming commonplace at major sporting events now to see a triumphant athlete raise his forefingers in a happy salute to the crowd. For some religious groups a similar gesture is employed to indicate that there is *one god*.

■ The middle finger has been looked upon as the obscene digit for more than 2,000 years. The Romans called it the digitus obscenus and employed it in a phallic gesture which was considered the ultimate insult. When the notorious Emperor Caligula offered his middle finger to be kissed by his subordinates, instead of the back of his hand, they were (as he intended) completely outraged. In recent times, Americans have referred to this phallic gesture simply as 'the finger', but in 1976 Nelson Rockefeller was rash enough to perform the gesture in the presence of a news cameraman, after which it became widely known as 'The Rockefeller Gesture'. In complete contrast, the third finger is an innocent digit which is used to symbolize love and marriage, as the 'ring finger'; while the crooked little finger, originally a sign of sexual independence, became an affectation of polite gentility.

◄ In Britain the most obscene gesture is the V-sign performed with the palm of the hand towards the gesturer. In other countries this simply means 'two' or 'victory', but in the British Isles, such messages would only be transmitted using the palm out V-sign. This is a local distinction which often confuses foreign visitors.

heal it and it eventually became known as the healing finger or the leech finger. In parts of Europe it is still used today as the only finger suitable for scratching the skin.

If there is any practical value in this superstition it is that of all the digits the ring finger is the least used and therefore probably the cleanest. The reason for its comparative inactivity is that its musculature renders it the least independent of all the digits. If you make a fist with your hand and then try to straighten out and bend back each digit, one at a time, it is only the ring finger that refuses to straighten itself out fully – or does so with great difficulty. If the finger on either side of it is straightened at the same time there is no problem, but on its own it feels too weak to make the movement. This meant that it was the least likely of the digits to have been touching anything harmful and was therefore the safest for medical use. It must also have meant that it was difficult to use as an efficient stirrer without keeping the other fingers pressed back out of the way by the thumb.

It was because of this lack of independence that the finger also became known as the ring finger. The ancient custom of placing the wedding ring on the third finger of the left hand was based on the idea that the wife was committing herself to be less independent, like the symbolic digit. The use of the left hand was based on the idea that this was the weaker, submissive hand, appropriate for the wife's subordinate role. It is perhaps surprising that we still use this finger today when in so many other respects we favour sexual equality.

Because of its role as the ring finger, it was also known to the Romans as the digitus annularis. In chiromancy it was dedicated to the sun and in palmistry to Apollo, the sun-god. In Islam it was given to Hassan; and for Christians it was the 'Amen Finger', based on the fact that gestures of blessing were made with the Thumb (The Father), the Forefinger (The Son), and the Middle Finger (The Holy Ghost), followed by the Amen of the ring finger.

The fifth digit: the Little Finger. In Latin this was known as the minimus or the auricularis, minimus because it is the smallest and auricularis because of its association with the ear. It is usually claimed that its title as the 'ear finger' is based on the fact that it is small enough to be useful in cleaning the ear, but this is probably a modern rationalization. In earlier days it was thought that by blocking the ears with the

Some hand gestures are elaborately stylized.

little fingers it was possible to increase the chances of a psychic experience, a prophetic vision, or some other supernatural event. Anyone who has attended a seance will probably have indulged in a modern survival of this superstition when joining hands in a circle. On such occasions the medium usually insists that it is the tips of the little fingers that are used to make contact with one's neighbours because this was the ancient way of forging a psychic link.

In America the popular name for this finger is the pinkie. The term was first used there by children in New York, but later spread to adults and to other cities. It is thought to have originated in Scotland, where children referred to anything small as a 'pinkie' and it is supposed to have been transported to the New World by Scottish settlers. However, the original name for New York was New Amsterdam and it may also be significant that the Dutch word for the little finger is *pinkje*. The children who employed the word often did so in a special rhyme which they chanted when making a solemn bargain. As

■ Because they are so visible and mobile during social encounters, the hands are a prime target for decorative embellishment. Dancers, like this girl from Thailand with her amazingly elongated finger-decorations, or this Indian performer from Madras with her conspicuously red-tipped fingers, increase the impact of their hand movements in a dramatic way.

■ The intricately painted hand, opposite, is also from India, but in this case is not that of a dancer but of a bride-to-be. Her hands, painted with a mixture of henna and clay, will retain this delicate tracery for several weeks, before it begins to fade and wear away.

The symbolism of the hands is often extended by decoration.

▲ *Finger-rings have been popular as a form of decoration since prehistoric times. To some they were protective, to others merely symbols of wealth and high status. The habit of putting a wedding ring on the third finger was said to stem from the idea that, since this is the digit with the least independent movement, it is the appropriate place to put a symbol which indicates the loss of independence inherent in marriage.*

they did so they interlocked their little fingers to make the bargain binding. This is yet another survival of the ancient psychic-link role of the little finger. In some European countries, when two people accidentally say the same word at precisely the same moment they shout 'Snap' and then interlock their little fingers. While doing this they are both permitted to make a silent wish, which will come true if nothing is said until their fingers are released. Once again, this reflects the ancient belief in the psychic power of the little fingers and their ability to transmit supernatural forces. The reason for saying 'Snap' also has to do with the digits, because it is a verbal substitute for the action of snapping the fingers, another act with superstitious origins. It used to be thought that the loud noise made by the snap of the forefinger against the thumb would scare away evil spirits (which is why it is rude to snap your fingers for attention), and this was thought necessary when two people uttered the same word simultaneously.

In a completely non-magical context, the crooking of the little finger when drinking from a cup or a glass has long been thought of as the height of genteel affectation. In origin, nothing could be further from the truth. Early religious paintings often show the little finger crooked away from the other digits, even when the female figure in question is not drinking. It is claimed that this was a sign that the real life models for the religious figures were girls with an unusual degree of sexual independence. This belief that an 'independent' little finger symbolized sexual freedom was the basis of a new fashion started by the members of the women's movement at the end of the last century. They deliberately crooked their little fingers when drinking to display their support for the idea of equal rights in sexual matters. Slowly spreading as a fashionable finger posture, this action gradually shed its original meaning and eventually lost its sexual significance, becoming merely 'the thing to do' when in company. From there it was on its way to becoming a sign of polite gentility, ending up with almost the opposite meaning to its original one.

Regional gestures employing an erected little finger have several different meanings, according to where you are. In Europe it can signal that someone is very thin or ill – as skinny and undernourished as a little finger. In obscene contexts, in the Mediterranean region, it has a more specific meaning, suggesting that someone has a very small penis – about the size of the little finger. On the other side of the world, in Bali, it signifies that something is bad. In Bali a gesture with the raised thumb means that something is good – just like our own 'thumbs up', but instead of having the thumbs down as the opposite of this action they use this other kind of 'opposite' – the erected little finger.

Up to this point, the actions of each digit have been considered separately. When they are used in various combinations a whole new range of gestures and signs is made possible. Some of these are universally employed and are performed unconsciously whenever we find ourselves in a certain mood. We make our hand into a grasping claw, a chopping edge, a tight fist or a spreading fan, according to the emotions that obsess us from moment to moment. We do not think of what we are doing in such cases and would find it hard to recall afterwards precisely what we did, but despite this the message of the

With care, the untrimmed fingernails will grow to extraordinary lengths.

■ In some cultures fingernails have been allowed to grow to amazing lengths to display that their owner does not need to carry out any manual work. In other places, excessive nail growth has had a religious significance, but today the prime motive appears to be to gain an entry in The Guinness Book of Records. In order to maintain something approaching a normal existence, these record-breakers concentrate only on the left hand, trimming the right hand for day-to-day use.

moving hands gets across to our companions well enough.

Finally, there are the fingernails – dead tissue growing from a living base at a rate of one millimetre every ten days. This rate of growth means that, uncut, we could grow our nails about one centimetre longer in 100 days, or an inch in about eight months. Many females in different epochs and in different cultures have done this as a sign that they are not required to undertake any form of manual labour. This high-status display is enhanced by the application of brightly coloured varnish to the nails to draw attention to the fact that *these* are hands which never have to toil. In ancient China both sexes of the nobility grew their nails long for this reason and painted them gold. Later, because this made their ordinary hand movements so cumbersome, they restricted the display to the little fingers only, clipping the others much shorter. Another solution was to have shorter nails for everyday use and then to clip on wildly exaggerated false nails for special occasions. Both these practices are found today in Europe. Many women use false nails which they glue on for social events and then remove for work. And young men in certain Mediterranean countries still employ the single, long little-finger nail to demonstrate that they do not have to do heavy rough work to earn a living. These long nails are sometimes used for rather painful, supposedly erotic nipping during courtship and love-making, as female tourists visiting the Mediterranean shores have sometimes discovered to their cost.

In some cultures the fingernails, which grow about four times as fast as toenails, have been allowed to grow to astonishing lengths. The world record for extreme fingernail length is held by an Indian, Shridhar Chillal of Poona, who managed to grow the nails of his left hand to a combined total length of 115½ inches and sported a thumbnail no less than 29½ inches long. This extraordinary achievement was the result of thirty years of devoted care and attention to his left hand, his right one being left free to deal with everyday matters.

■ There are literally hundreds of different regional hand gestures, each with its special local meaning. These can prove extremely confusing as we travel around the world and, as yet, nobody has produced a global dictionary of hand gestures to assist the traveller. Fortunately, many of these actions are easily enough understood, like the 'tiger claws' gesture of this small boy in a Paris street.

THE CHEST

■ When our ancient ancestors switched to hunting as a means of survival, new pressures came to bear on the human body. The males who set off on the chase had to develop improved respiration. If they ran out of breath they ran out of food. Compared with other monkeys and apes they had to become big-chested. To house the larger lungs the bone-cage formed by the backbone, ribs and sternum had to become more barrel-shaped. It grew in both length and breadth. The male chest became an athlete's chest.

The female developed in a different way. Hampered by pregnancies and infants, she was less mobile. Her chest did not enlarge like the male's. It developed in another direction, the rib-cage remaining small but the breasts swelling to a pair of soft hemispheres. These enlarged breasts had two biological functions, one parental and the other sexual. Parentally they act as gigantic sweat glands producing the modified sweat we call milk. The glandular tissues that produce the milk become enlarged during pregnancy, making the breasts slightly bigger than usual. The blood vessels serving these tissues become much more conspicuous on the breast surfaces. As the milk forms it passes along ducts towards special storage spaces called sinuses. These are positioned in the centre of the breast behind the dark-brown areolar patches that surround the nipples. From these

■ *The broad, flat and sometimes hairy chest of the human male contrasts strikingly with that of the human female. This difference is due to the division of labour between the sexes which developed during our long evolution as a hunting species. The males had to become better athletes and this required more efficient breathing. Bigger lungs demanded larger, heavier chests. This explains why a male 'swells his chest with pride'. By so doing he is exaggerating one of the body-badges of his masculinity.*

■ Breast-feeding is easier for an infant monkey than for an infant human: there is maternal fur to cling to, and long nipples protruding from a conveniently flat chest. The human mother's short-nippled breast is so rounded that its shape sometimes threatens to smother the hungry face pressed against it.

■ The chest of the adult human female is unusual because it displays swollen breasts even when no lactation is taking place. This is unique among primates. It indicates that the hemispherical shape of the breasts is connected with sexual display and not parental feeding. When a female does start lactating, and her breasts become heavy with milk, they increase in size by about one-third. This means that two-thirds of their size – enough to give them their hemispherical shape – is non-parental.

sinuses there are some 15 to 20 tubes, the lactiferous ducts, leading to each nipple.

When a baby sucks it takes the whole of the areolar patch and the nipple into its mouth, squeezing the brown skin with its gums, and in this way squirts the milk out of the nipple. If it takes only the nipple into its mouth it has a problem, because squeezing the nipple alone does not produce the desired milk. It may respond to this frustration by chewing on the nipple, which does little good either to mother or offspring. An inexperienced mother soon finds that she can avoid the pain caused by these hungry attentions by squeezing more of her breast into the baby's mouth.

The areolar patch surrounding the nipple is an intriguing ana-

▲ Because of their sexual rather than parental shape the female breasts are frequently displayed in erotic dancing. The suggestion that in such cases the watching males' response is infantile is one of the many errors of psychoanalytical thinking.

◀ The male chest muscles are much heavier and stronger than those of the female and attention is drawn to this fact in warlike chest displays by males who inflate their lungs and sometimes, like gorillas, beat their out-thrust chests with hands or fists.

tomical detail of the human species. In virgin females and those who have yet to become mothers it is a pinkish colour, but during pregnancy it changes. About two months after conception it starts to grow larger and becomes much darker. By the time lactation has started it is usually a darkish brown colour and later, when the baby is weaned, it never quite returns to its original virginal pink. In function these areolar patches appear to be protective. They are full of specialized glands that secrete a strange fatty substance. To the naked eye the glands look like 'goose pimples' on the pigmented skin. During the breast-feeding phase they become much enlarged, and are then called Montgomery's tubercles. Their secretions help to protect the

◄ *The striking similarity between human female breasts and buttocks has sometimes been exploited by studio photographers. The twin-hemisphere echo which they create reflects, in reality, not an imaginative leap by the artist but the result of a long, slow evolutionary development that has provided the human female with a powerful frontal sexual signal.*

skin of the nipple and its surrounding skin – a form of biological 'skin care' much needed by the lovingly abused surface of the breast.

The milk produced by the female breasts contains proteins, carbohydrate, fat, cholesterol, calcium, phosphorus, potassium, sodium, magnesium, iron and vitamins. It also contains various antibodies which may make the infant more disease-resistant. Cow's milk is a fair substitute for mother's milk, but its phosphorus level is rather high and may interfere with the intake of calcium and magnesium by human infants. Also, some babies may have allergic reactions to bovine proteins. Wisely, more mothers are breast-feeding nowadays – and there is the added bonus that it forms a much stronger bond of love between mother and baby.

While mother's milk is ideal for a growing baby it has to be said that the shape of her breast is far from perfect for the task of breast-feeding. The teat of a milk bottle is much more suitably shaped for delivering milk into the infant mouth than the real nipple on the mother's breast. If this appears to be an evolutionary flaw, it must be remembered that the female breast has a dual role – parental and sexual – and it is the sexual factor that causes the problem here. To understand why this should be it helps to cast a sidelong glance at the breasts of our close relatives, the monkeys and apes.

In all other primate species the females are flat-chested when not lactating. When they are breast-feeding the region around the nipples becomes somewhat swollen with milk, but even then it is rare to find anything approaching the hemispherical shape of the human female breast. In the case of those few that do approach the human shape when they have a particularly generous milk supply the swellings disappear after lactation is complete. The 'breasts' of monkeys and apes are purely parental.

The breasts of the human female are different. Although they increase in size slightly when full of milk they remain protuberant and firmly shaped throughout the period of young adulthood regardless of parental consideration. Even a nun has protuberant breasts though they remain unused through her lifetime.

An examination of the anatomy of the breast reveals that most of its bulk is made up of fat tissue, while only a small part is glandular

By mimicking the shape of the buttocks, the female breasts echo their primeval signal.

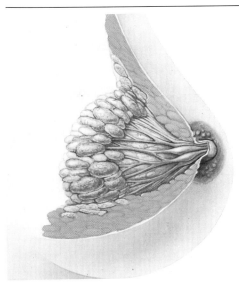

■ *The larger part of the human female breasts consists of fatty tissue which is concerned, not with milk supply, but with giving the breasts their rounded shape. Females with exceptionally well developed busts on otherwise slender figures transmit supernormal sexual signals which attract males at a primeval level; and it has been pointed out that breasts unsupported by tight, modern clothing display a mobility during locomotion similar to that of buttocks, thus helping to improve their rump-echo sexuality.*

▶ *The ancient Diana or Artemis of Ephesus has nearly always been described as a 'Many-breasted Earth Goddess', but this simply reflects the ease with which men see breast-symbols in spheroid shapes. In reality, it is now known that she was a Virgin Mother who was ritually adorned in castrated bull's testicles as part of her mystical, symbolic fertilization.*

tissue concerned with milk production. The hemispherical shape of the breasts is not a parental development. It is concerned instead with sexual signalling. This means that suggestions that men's interest in women's breasts is 'infantile' or 'regressive' is unfounded. The male responding to the prominent breasts of a virgin or non-lactating female is reacting to a primeval sex signal of the human species.

The origin of paired hemispheres of human female sexual signal is not hard to find. The females of all other primates display their sexual signals backwards from the rump region as they walk about on all fours. Their sexual swellings are key stimuli which excite their males. The rump signals of a human female consist of unique paired hemispheres, the buttocks. These can act as powerful erotic signals when she is seen from behind, but she does not go around on all fours like other species, with her frontal region hidden from view. She stands upright and is encountered frontally in most social contexts. When she stands face-to-face with a male her rump-signals are concealed from view, but the evolution of a pair of mimic-buttocks on her chest enables her to continue to transmit the primeval sexual signal without turning her back on her companion.

So important was this sexual element in breast development that it actually began to interfere with the primary parental function. The breasts became so bulgingly hemispherical in their efforts to mimic the buttocks that they made it difficult for a baby to get at the nipples. In other species the female nipples are elongated and the baby monkey or ape has no difficulty whatever in taking the long teat into its mouth and sucking away at the milk supply. But the human infant of a well-rounded mother may be almost suffocated by the great curve of flesh that surrounds the rather modest human nipple. Such mothers have to take precautions that would never be needed in any other species. Dr Spock advises: 'At times you may need to put a finger on the breast to give him breathing space for his nose.' Another baby book comments: 'It may surprise you that he takes the brown area around the nipple in his mouth as well. All you need to do is to make sure that he can breathe. In his eagerness he may obstruct his nostrils with breast tissue or with his own upper lip.' Cautions such as these leave no doubt about the dual role of the human breasts.

Women who have rather small breasts often worry that they will not be able to breast-feed. Ironically they may well be able to breast-feed more efficiently than their well-rounded friends. This is because they have less of the fat tissue that gives the breasts their sexual hemispherical shape but which has little to do with the milk supply. Once they become pregnant their glandular tissue will increase in size, as with all expectant mothers, but they will not have such bulbous breasts as their heavier companions, and their babies will find sucking much easier and less suffocating.

In their sexual role the female breasts operate first as visual stimuli and then as tactile ones. Even at a great distance they are normally sufficient to distinguish the silhouette of the adult female from that of the male. At a closer range this crude gender signal gives way to a more subtle age-indicator. The shape of the breast changes gradually from the age of puberty to old age. This slow alteration in mammary profile can be simplified into the 'seven ages of the female breast' as follows: (1) The nipple breast of childhood. Only the nipple

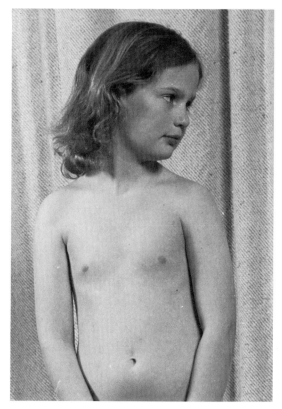

■ *The breasts of other primates rise and fall with each cycle of pregnancy and lactation. The breasts of the human female, by contrast, rise at puberty and remain swollen throughout her adult life, until they shrink and sag in very old age. They pass through seven stages, five of which are seen here. From left to right: the nipple breast of childhood; the firm breast of young adulthood; the full breast of motherhood (and, in the background, the shrunken breast of old age); and the pendulous breast typical of middle age (but seen here in a younger female of heavy build).*

▶ *Two opposing ways of drawing attention to the breasts: by protecting or covering them with exaggerated mock-modesty, and by brazenly raising them and emphasizing their hemispherical shape with the help of the hands. During the history of fashion, clothing has been designed on both these principles – in one epoch exposing cleavage and in another submerging the torso right up to the neck, but without destroying the outline of the breast.*

The life-cycle of the female breasts.

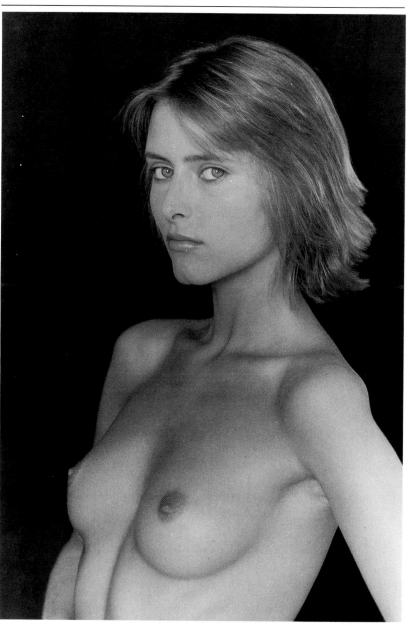

is elevated in this prepubertal stage. (2) The breast-bud of puberty. At the very start of the reproductive phase, when menstruation begins and the genitals start to sprout pubic hair, the region around the nipple starts to swell. (3) The pointed breast of adolescence. As the teenage years pass there is a further slight increase in breast size. At this stage both the nipple and the areolar patch project above the breast, creating a more-pointed conical shape. (4) The firm breast of young adulthood. The ideal physical age for the human animal is 25. This is the stage at which the body is at its peak of condition and all growth processes have been completed. During the twenties the female breast fills out to its most rounded hemispherical condition. Although it is larger its weight has not yet started to make it droop. (5) The full breast of motherhood. With maternity and the sudden additional bulk of enlarged glandular tissue, the milk-laden breast balloons out and starts to turn downwards on the chest. The lower

margin of the breast overlaps the chest skin to create a hidden fold. (6) The sagging breast of middle age. As the reproductive phase of adulthood nears its end, the breasts hang further down on the chest even though they have lost the fullness of the lactation stage. (7) The pendulous breast of old age. With advancing old age the general shrinkage of the body leads to a flattening of the breasts which remain hanging down on the chest but with increasingly wrinkled skin.

These are the typical stages in breast aging, but there are many variations on the theme. In thinner women the process tends to slow down to some extent, while in fatter ones it speeds up. Cosmetic surgery can prop up breasts and artificially extend the firmness of the young adult stage. Costume supports such as corsets and bras can give the same impression, providing the breasts are not directly visible. In a variety of ways over the years women have sought to prolong the impression of firmly protruding hemispherical breasts in order to extend the period over which they are able to transmit the primeval female chest signal of the human species.

Sometimes the mood of society has demanded that the sexuality of the female bust be suppressed. Puritans achieved this by forcing young women to wear tight bodices which flattened the breasts and gave an innocent childlike contour even to mature adults. In seventeenth century Spain young ladies suffered even greater indignity, having their swelling breasts flattened by lead plates which were pressed tightly on to their chests in an attempt to prevent nature from taking its protuberant course. Such cruel impositions only serve to underline the intense sexual significance of the hemispherical shape of the breast. For society to go to such lengths to negate it it must indeed be powerful.

Happily most societies have been prepared to accept the covering rather than the crushing of the breast as a sufficient expression of modesty. In such instances the mere removal of this covering has acted as a major erotic stimulus. This has been exploited by artists and photographers in many different ways. For the artists it is easy enough to produce the perfect breast; they can invent any breast shape they like within reason. If they stray too far from nature the primeval signal becomes distorted and its impact is lost. But if the basic hemispherical shape is made slightly more hemispherical than usual it is possible to create a super-breast which is perhaps even more stimulating that the real thing.

The photographer has a more difficult task. Confined to real breasts he can only hope to improve their shape by special lighting or by arranging his models in postures that enhance the hemispherical signals. He can of course ensure that his models are both metaphorically and literally at their peak of breast development. To capture a super-breast he needs a model whose adolescent breasts have reached maximum development – just before their increasing weight starts to make them droop. He finds a conflict of forces here because the increase in size which gives the full hemispherical shape inevitably also adds to the weight and starts to pull the breast downwards. There is only one point in a female's life when her breasts have maximum protrusion with minimum droop and that is the moment when his camera shutter must click if he is to produce the most erotic images.

It is interesting that expert photographers who work for glossy

■ *Because it transmits a very basic, primeval sexual signal, the female breast-shape has an irrestistible tactile appeal for the male. There are many occasions when his desire to reach out and touch the curve of the breast is frustrated by social restrictions, but opportunities to satisfy this ancient urge do sometimes arise. Occasionally females go out of their way to invite such attention.*

magazines specializing in erotic pictures find that there is only one kind of girl with the super-breasts they seek. Her age is slightly younger than one might expect, namely the late teens, and her breasts have grown to full adult size slightly earlier than the average: they exhibit the perfect roundedness required, but still retain the firmness of extreme youth. This special combination provides the kind of images of which centrefolds and men's-magazine fortunes are made.

Once the visual signals of the female breast – and her other physical and mental charms – have attracted a male partner and sexual contact has begun, the tactile qualities of the breasts come into play. In pre-copulatory sequences there is often a great deal of oral and manual caressing of the breasts by the male. This excites the male even more than the female and it is possible that a special additional stimulus is operating here. It was mentioned earlier that the brown patches of skin around the nipples contain glands that secrete a fatty substance during lactation. This is claimed to be a soothing lubricant for the over-worked skin of the nipple region, and there is no reason to doubt this. But the fact that the glands of the areolar region are in origin apocrine glands suggests that during sexual activity the nipple zone of the breasts of the female may actually transmit scent signals to the male nose. Apocrine glands are the ones responsible for the special sexual fragrance of the armpits and the genital regions, and although males are not consciously aware of the erotic odours these glands produce their secretions do make a massive unconscious impact which aids sexual arousal. The areolar glands may well be part of this odour-signalling system, which may explain why males exploring their partners' bodies spend so much time nosing around in the mammary zone.

As sexual arousal mounts, the female chest undergoes several marked changes. The nipples become erect, increasing in length by up to a centimetre. The breasts themselves become engorged with blood, increasing their overall size by up to 25 per cent. This turgidity has the effect of making their whole surface more sensitive and responsive to the body-to-body clasping of the mating pair.

With the approach of orgasm two further changes occur. The areolar patches become tumescent and swell so much that they start to mask the nipple, giving the false impression that the strongly aroused female is actually losing her nipple erection. There also appears a strange measles-like rash over the surface of the breasts and elsewhere on the chest. This 'sex flush' was observed to occur in 75 per cent of women who were the subjects of a detailed sex investigation. It is far less common in men but was seen in 25 per cent of those taking part in the same investigation. It is most likely to occur during the moments immediately before orgasm in both sexes. In females, however, it may sometimes appear quite a while before orgasm is reached, whereas in males it never appears until the last moment. Although it is not possible to develop this rash without experiencing intense sexual arousal, the converse is not true. Many individuals of both sexes never display the sex flush despite vigorous sex lives full of orgasmic experiences. Why people should differ in this way is not known. One important factor that favours the rash is a hot atmosphere. In cool conditions individuals who might otherwise display the sex flush do not do so. When it is very hot, on the other hand, the

Chest decoration is usually restricted to painting or the wearing of pendant ornaments.

■ *In tribal societies, breast mutilation is rare for the obvious reason that, without the possibility of bottle-feeding, the breasts must be kept in perfect working order for suckling babies. Breast decoration is therefore usually restricted to painting and to letting ornamental objects hang over and around them. The same is true of the more exotic displays in urban societies, although recently certain females have started to break this rule in a bizarre way, having their nipples pierced for the attachment of rings and chains. In some instances, even males have followed suit. These developments, although comparatively minor, reflect the bondage themes of the world of sado-masochism and the recent punk rebellion against all forms of bodily propriety. The vulnerability of such individuals is obvious, since they could be led, not by the nose like a bull, but by this much more sensitive part of the anatomy, should anyone wish to control them. The wearing of these nipple rings therefore either implies the desire to be led around like a slave, or the complete and ostentatious lack of fear that anyone would dare to do such a thing.*

rash may extend far beyond the primary chest region, covering an area that ranges from the forehead to the thighs.

One of the facts that we all take for granted is that human females have only two breasts, but this is not always the case. About one in every two hundred women has more than two. There is nothing sinister in this and the additional breasts are usually non-functional. Sometimes they are little more than additional nipples, sometimes small breast-buds without nipples. They are remnants of our very ancient ancestry: like most other mammal species, our remote forebears had several pairs with which they could feed a whole litter of offspring. When our litter size fell to one, or occasionally two, the number of nipples diminished in parallel.

Several famous women have possessed more than two breasts. Julia, the mother of the Roman emperor Alexander Severus, was many breasted and was given the name Julia Mamaea as a result.

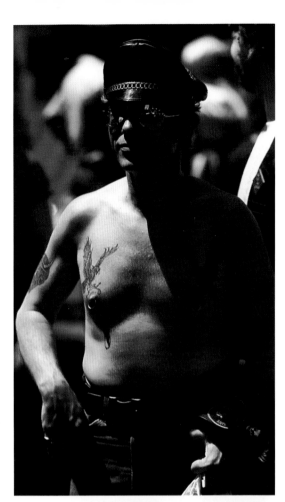

More surprisingly, close examination reveals that the famous statue the Venus de Milo, in the Louvre, displays three breasts. This is usually overlooked because the third breast has no nipple and is little more than a small breast-bud. It is situated above the right breast, near the armpit. Henry VIII's unfortunate wife Ann Boleyn was also said to possess a third breast – a claim faithfully recorded in books on medical abnormalities. In this case, however, the alleged third breast may well have been a 'witchcraft' slur. Witches were once believed to have extra nipples with which they suckled their familiars, and women who were thought to be guilty of witchcraft were sometimes searched for telltale signs of their evil ways. Pious Christian witch-hunters would diligently search the most private crevices of a suspected witch for a hidden nipple. A wart or a large mole and sometimes even a slightly enlarged clitoris would be enough to have the wretched owner burned to death at the stake. Rumours of Ann Boleyn's third breast may have been deliberately spread at her death to suggest that she was evil and deserved to die.

The most famous multi-breasted figure in history is the Diana – or Artemis – of Ephesus. Her ample sculptured bosom displays several rows of breasts crowded together. Some versions of the statue show more than twenty breasts. Or do they? A closer look reveals that none of these breasts has a nipple or an areolar patch. They are all 'blind breasts'. Recently the cult of this ancient Anatolian mother goddess has been examined more carefully and an entirely new interpretation has emerged. To understate the case, the Diana's chest is a less friendly place than has long been supposed. It seems that the goddess's arch-priest had to be a eunuch: in order to serve her he had to castrate himself and bury his testicles near her altar. Inscriptions have been found which reveal that after a time bulls were substituted for priests at the castration ceremonies. Their huge testicles were removed and preserved in scented oils, then ceremonially hung on the chest of the sacred statue. The original statue was made of wood, but copies of it were made in stone, with the cluster of sacrificed bulls' testicles shown in place. It was a study of inaccurate stone copies that gave rise to the long-standing error that the Great Mother was many-breasted. The reason for covering the chest of the goddess with the testicles was that the millions of spermatozoa contained in them were thought to fertilize her. This was achieved in such a way that she could become a mother while remaining a virgin, a theme that was to reappear in connexion with the birth of Christ.

A breast myth of an entirely different kind surrounds the ancient nation of female warriors known as the Amazons. It is doubtful whether they ever really existed but according to early writers they were a fearsome all-female community forever attacking neighbouring peoples with their bows and arrows. In order to release their arrows more efficiently they were said by some to burn off the right breast of all girls at puberty. Others had it that the offending breast was cut off. Unfortunately for these stories all the ancient works of art showing these ferocious females depict them with two sound breasts. If they existed at all it is more likely that they wore a one-sided leather tunic which flattened the right breast during battle. The name Amazon means literally 'without breast' (a-mazos).

Curiously in recent years westerners have started to mutilate their

■ *The act of placing a hand on one's chest is extremely ancient, going back to classical Greece and beyond. It has been used as a sign of loyalty and also as a way of swearing an oath. For Greek slaves, it was a gesture of obedience, signifying that they were awaiting the command of their master. Today it is most commonly observed in the United States, where it is employed by non-military individuals during the playing of the national anthem, in place of the usual military salute. Its origin is obvious enough – it symbolizes placing the hand on the heart, but it is less obvious what, in this context, the heart itself stands for. In modern times we think of the heart as the symbol of emotions and feelings, but this was not the case in ancient times, when the Hand-on-heart originated. Then, the heart was considered to be the very essence of the person – his intelligence, and the centre of his being. It was this that the ancients were symbolically touching when they stood hand-on-heart. The brain, in those early days, was seen simply as the instrument of the heart's intelligence. Because of this origin, which still influences us today, placing a hand on someone else's chest is a rather intimate action, usually performed only between lovers or very old friends.*

The origin of the Hand-on-heart gesture.

breasts for erotic and decorative purposes. Cases are rare, but sufficiently widespread to alarm sociologists, one of whom declared that the new fashion of 'erotic body piercing of nipples, navel and labia, and the insertion of chains, jewellery etc' could easily forestall sensible legislation to outlaw the African custom of female circumcision. Modern nipple-piercing is probably a variant of the punk rock fashion of piercing the skin with safety pins and then hanging chains on them. Essentially this is part of the bondage syndrome from the world of exotic sexual practices. Among tribal societies mutilation of the breast is extremely rare for the obvious reason that it interferes with breast-feeding – a serious drawback where bottle-feeding is not an available alternative.

When it comes to chest gestures, there are two main symbolic elements involved. The chest region is either used as a representation of the 'self' or (in females) as a sexual zone.

The chest as 'self' is seen whenever a speaker wishes to emphasize the concept of 'I' or 'me'. As he uses these words he touches or taps his chest with his fingers. In moments of great happiness a chest-hugging gesture may be performed in which the person concerned embraces his own chest with his arms.

Puffing out the chest or beating it with hands or fists is a masculine display common to many cultures. It signifies assertion and it too uses the chest region as if it stood for the whole person. Covering up this 'self' region can act as the opposite signal. In the Orient folding the arms across the chest is a sign of humility which accompanies bowing. In ancient Greece the left hand held to the chest was a slave's sign of obedience; and among Arabs, touching the chest, along with the mouth and the forehead, is a polite form of greeting.

Mourners in ancient times used to bare their chests or beat them in grief. In early literature a mother was said to bare her breasts as a display of sorrow over a departing child, reminding him of how she had once fed and cared for him as a baby.

Among the sexual chest signals are various forms of breast-cupping with the hands, chest protrusions, and dance movements that shake or emphasize the shape of the breasts. All these draw attention to the sexual hemispheres of the female. The most extreme is the famous tassel dance of old-style burlesque shows in which the performers rotated both breasts in the same direction and then in opposite directions, the tassels following suit.

A final word is necessary to clear up a misunderstanding about ancient figures which are shown squeezing their breasts with their hands. These have always been called Mother Goddess images and it has been supposed that they were emphasizing their breasts by cupping them erotically. We now know that this was not the case. These figures, usually found in ancient tombs, were mourner figures. In early times females performed a mourning ritual that involved both beating and squeezing their breasts. A side-effect of this, if they were lactating at the time, was that the squeezing sometimes squirted milk from their nipples in long jets. It is possible that this action eventually became incorporated into certain rituals. Anthropologists have found, to their surprise, that in certain remote tribal societies lactating females react in a similar way to sudden shock, grabbing at their breasts in panic and squirting milk several feet in front of them.

THE BACK

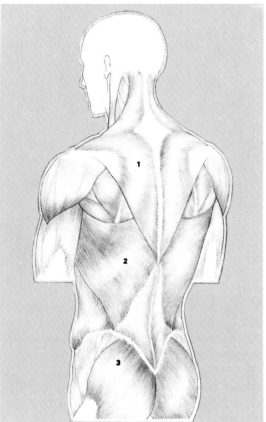

■ The human back is the hardest-working yet least-known region of the body. Ever since our remote ancestors stood up on their hind legs, our back muscles have been forced to work overtime, and it is a rare individual who does not, at some time in life, suffer from a nagging backache. For most of us this is the only time we ever really pause to consider our backs as a separate part of our anatomy. At other moments it is a case of 'out of sight, out of mind', and there are few of us who could identify ourselves in a backs-only line-up.

If we did take the trouble to have a closer look at our long-suffering backs we would find a brilliantly assembled set of muscles and bones with the twin functions of body support and spinal-cord protection. The cord itself, which is about 18 inches long and just over two-fifths of an inch in diameter, certainly needs protection. If anything serious happens to it, it is time to buy a wheelchair. The back securely encases it, first in a triple-layered sheath, second in a shock-absorbing fluid, and third in a hard, blow-resistant casing we call the backbone. In reality, of course, there is no such thing as the backbone – there are 33 bones, in a long series. These, the vertebrae, are of five different types: The cervical or neck vertebrae have an amazing degree of mobility, permitting all the various head movements so vital to scanning our world and protecting our face. There are seven of these bones. The thoracic or chest vertebrae are much less mobile because their main job is acting as an anchor for the ribs. There are twelve of these vertebrae. The lumbar, or 'loin', vertebrae, the heaviest and stoutest of all, have the task of carrying the greater part of the body's weight. It is here that the dreaded backache is most likely to strike. There are five of these bones. Then come the sacral and coccygeal vertebrae.

The sacral, or 'sacred' vertebrae are fused together beneath the

■ *When human beings adopted a vertical posture they put heavy demands on the muscles of the back. The* trapezius *(1) of the upper back, the* dorsal *(2) of the middle back and the* gluteal *(3) of the lower back, all suffer from back strain if we underestimate their daily task in keeping us upright and forceful in our movements.*

◄ *Female opinion is divided on the sexual appeal of a hairy male back. Some see it as erotic because it is an exaggerated version of the naturally hairier male body. Viewed in this way it becomes a supernormal gender signal. Others see it verging on the apelike and anti-sexual because it looks less human. They also stress that they prefer the touch of a sensuously smooth skin when they embrace their males.*

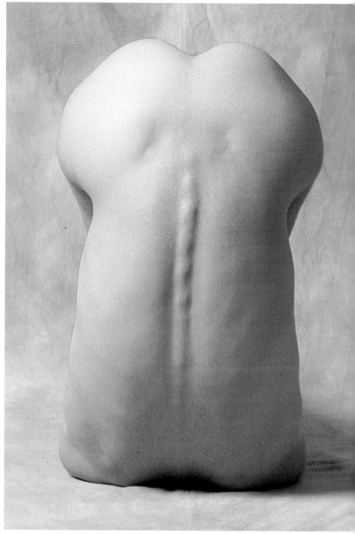

■ Unless we take special care to develop our back muscles we are in constant danger of straining them when we undertake vigorous lifting or climbing actions. Exposing our backs to the heat of the sun helps to improve blood circulation to the muscles and expert massage soothes the muscle tension from which so many urbanites suffer today.

■ On a well-padded female back a pair of sacral dimples is visible just above the buttocks. Less conspicuous on the skinny backs in favour today, these were once considered a feature of great beauty.

■ *The exposure of the back region, while keeping the rest of the body covered, has acted as a restrained display of eroticism on many occasions in the history of fashion. It succeeds because it permits the display of a large area of naked skin without the blatant offering of any specific sexual details. In Japan, especially, the back has been looked upon as a zone of intense erotic interest.*

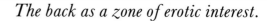

The back as a zone of erotic interest.

lumbar region to form the curved sacrum. This consists of five vertebrae but they now act as one. It may seem curious that this triangular bone at the base of the spine should be thought of as sacred, but in occult circles it is designated as the most important bone in the body and plays a special role in rituals involving divination by the use of 'holy bones'. It was also considered to be the bone containing the body's immortal spirit. To most people, however, there is something strangely perverse about locating the 'soul' at the bottom of the back. Perhaps this was the point, because it is the sacral bone which is ceremonially kissed at the witches' sabbath.

The coccygeal vertebrae are the smallest and lowest of the bones of the back. They are also fused together, to form the coccyx: all that remains of our primate tail. The naming of this tiny, pointed bone is even stranger than in the case of the sacrum, for the word 'coccyx' means 'cuckoo'. You may well wonder what possible connection there might be between our remnant of a tail and a bird like a cuckoo. The answer lies in the bone's unusual shape, which early anatomists claimed was reminiscent of a cuckoo's bill. Some of our bodyparts have acquired their names in delightfully eccentric ways.

The muscle system of the back is extremely complex, but consists of three main units: the trapezius of the upper back, the dorsal muscles of the middle back, and the gluteus muscles of the lower back. Most cases of backache are caused by straining these muscles in some way. Omitting special medical problems, we suffer from back pains for one main reason – lack of exercise in our civilized, urban environment. If the back muscles grow weak from disuse, they can easily fall prey to misuse. We misuse them by poor posture, by sudden, unaccustomed violent action and by tension.

Poor posture arises from certain work-styles, where the body is asked to hold a particular position for hours on end. It also becomes a problem during the increasingly lengthy leisure time of the western world, where soft furniture has spread into every home. During the many hours spent watching television, talking or reading, the under-exercised urbanite snuggles into his soft chair or soft bed for comfort, like an infant seeking the security of its mother's body. Mentally, this embracing, gentle furniture creates a sense of safety and calm, but physically it is putting heavy demands on the back muscles, which struggle valiantly to keep the spine – literally – in good shape. This is especially punishing when the slumped or sprawled or curled figure on the soft surface is overweight. Pregnant women have come to expect backache as an almost inevitable hazard of carrying the weight of a child, but pot-bellied males, who carry an almost identical load in the same region, are often surprised when they start to suffer from similar symptoms.

Picking up heavy objects by bending forward and employing the back like a crane is another classic piece of misuse which often tests the human body beyond endurance. Although there may have been little risk attached to this type of activity when a prehistoric hunter lifted up the carcass of a fresh kill and slung it over his shoulder, there is considerable danger when the desk-bound commuter attempts a modern counterpart of this manoeuvre.

Mental tension is another way of testing-to-destruction the human back. Body tensions caused by mental anguish and anxiety can lead to

64ᵉ Année N° 20   Le Numéro : 2 francs   Samedi 15 Mai 1926

LA VIE PARISIENNE

....EN MAI, FAIS CE QU'IL TE PLAIT

prolonged strain on the back muscles. Eventually they begin to ache and this causes even greater anguish, which causes...and so on, until a doctor is needed. This process often takes place almost unnoticed at first and may be initiated by emotional problems that preoccupy the brain so much that side-effects are ignored until it is too late. One of the major causes of backache, it appears, is sexual frustration, and greatly increased sexual activity is sometimes suggested as a cure.

Turning from pain to pleasure, the back has occasionally figured powerfully in the world of erotic imagery. The Japanese, particularly, are fond of this region in terms of sexual appeal. The kimono is cut away from the back of the neck to a precise degree, according to the status of the wearer. If she is a married woman, the alluring line of her spine is barely suggested, but if she is a geisha, the kimono is cut so that it stands away from the back of the neck. When she kneels down in front of her male companion, he is able to glimpse the whole length of her back, tantalizingly revealed by the stiffness of her costume.

In the west, costume designers have, from time to time, placed erotic emphasis on the back. If dresses are being worn with high fronts, then the interest may be shifted around to the rear, with low-cut lines exposing most of the upper back. The extreme version of this was Ungaro's famous jumpsuit of 1967 in which the whole of the back was exposed, right down to the top of the buttock-cleft. This cleft then produced a 'cleavage-echo' relating the lower back to the female's upper-chest region. It also gave the wearer the possibility of displaying her sacral dimples and her 'lozenge of Michaelis'. These are female back details which in the past have often aroused male ardour to the level of passionate obsession. They are less evident on the slender figures so popular today but, when the more voluptuous female shapes were in vogue, they became favourite talking-points among the more sophisticated libertines. The two dimples, which appear as small indentations on either side of the base of the spine just above the buttock region, are present in both sexes, but are much more distinct on the female back because of the fatty deposits in that area. In males they are so poorly differentiated that they are only visible at all in 18-25 per cent of cases. The classical world was fascinated by these female dimples and ancient poets sang their praises. Greek sculptors also paid loving attention to them. It is possible that the sexual appeal of female dimples on the cheeks of the face owes something to the presence of these other dimples close to the cheeks of the buttocks. The lozenge, or rhomboid, of Michaelis is a diamond-shaped zone situated between the sacral dimples, which was also a focus of erotic interest in earlier days.

In early symbolism, the back has little part to play, except as the housing of the spine. The spine itself was seen as a replica of the primal cosmic tree, reaching up to the heaven of the brain. Macedonians believed that, as a corpse rotted, its backbone turned into a snake. Other interpretations of the human spine saw it as a road, a ladder or a rod. In medieval times, the 'essence' of the spine was held to be unusually beneficial, and anyone who had more than his fair share of backbone was considered to be full of good luck. For this reason it was thought to bring good fortune to touch the hump of a hunchback. This belief still lives on in parts of the Mediterranean where small, plastic talismans can be purchased depicting a smiling hunchback in

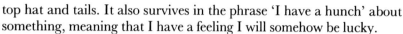

The cultivation of the female back muscles.

▶ *The new phenomenon of female body-building has demonstrated just how far it is possible to develop the muscles of the weaker female back. With specialized training, almost any female can strengthen her back far beyond the condition of the average, desk-bound, urban male.*

top hat and tails. It also survives in the phrase 'I have a hunch' about something, meaning that I have a feeling I will somehow be lucky.

The back is not one of the more expressive parts of the human body. Even the expression 'to get my back up' is based not on a human posture but on the arching of the back of an angry cat. We can, however, bend, stiffen, arch, slump, or wriggle the back as our moods change, and champion musclemen can even ripple their backs.

Bending the back forward, which for some elderly people becomes a chronic and permanent posture when walking, is an essential part of the subordinate acts of bowing, kneeling, kowtowing and prostrating. The vital element of all these actions is the lowering of the body to match the low status of the performer. Today we do little more than incline the head forward, but in earlier days the movement had to be extreme enough to expose the whole of the lowered back to the dominant individual. This was, in fact, the only way in which you could show your back to him. To turn your back on him when you were standing upright was considered an unforgivable rudeness. The point here is that showing your back to someone when you are in a deeply submissive posture renders you not merely lowly but also totally vulnerable, whereas turning your back on someone is an active movement of rejection. For this reason it was necessary for subordinates to vacate the presence of a Great One by walking out backwards from the room or royal chamber. Even today there is a remnant of this formal procedure, observable at a crowded party when someone twists his head and says 'excuse my back' to a friend who, in the crush, has come face to face with it. And sharply turning your back on someone to whom you have just been introduced remains to this day a major insult.

If turning the back is rude because it deliberately ignores a companion, stiffening the back is threatening because it suggests a bodily preparation for violent action. For this reason, the military are specially trained in back-stiffening so that, even when they are relaxed and at ease, they appear a little more aggressive than the average citizen. Stiffening the back also has the effect of slightly increasing the overall body height, a change that aids a display of dominance. Slumping the back, which occurs with depression (and incidentally gives the condition its name) transmits signals of loss of dominance by lowering the body slightly – almost as though one were observing an incipient bow of subordination.

Arching and wriggling the back are more sexual movements, the arching helping to protrude the buttocks of the female (whose back is naturally more arched than the male's even when at rest), and the

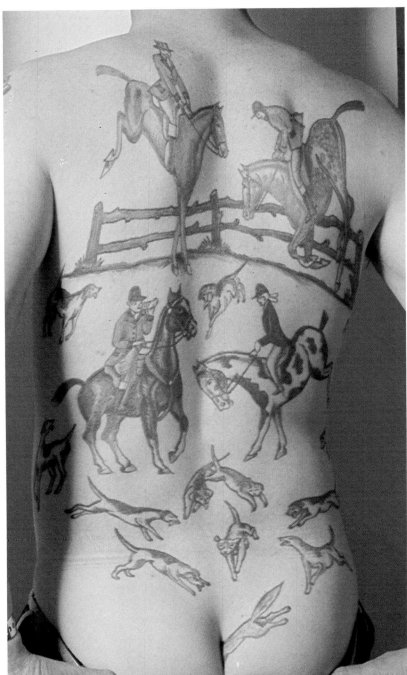

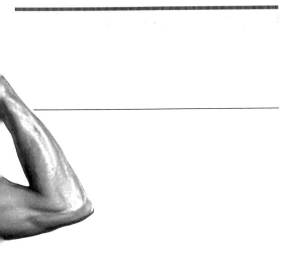

■ *Because of its large, flat expanse, the back is one of the most popular areas for detailed decoration such as tattooing. Its one drawback is that it is difficult for owners to appreciate the display since it can only be studied in photographs or glimpsed in mirrors. For the tattooee, it is one of the great ironies of body-design that the best 'canvas' is in the worst position. These back-tattoos have been described as the 'ultimate form of traditional anti-fashion'. They make nonsense of a fashion industry which relies for its economic survival on constant change and the cycling of styles.*

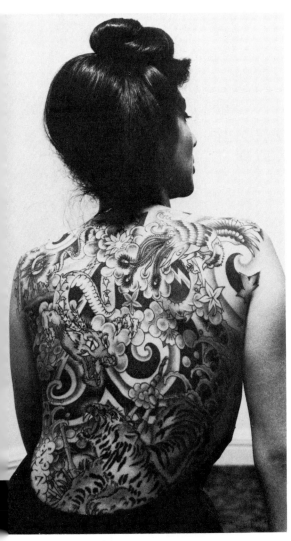

wriggling helping to accentuate the feminine curves and contours. The contours of the back are, of course, strikingly different in the human male and the human female, with the lower back being wider in the female and the upper back wider in the male.

When we make contact with our own backs there are several characteristic actions. The simplest is standing or walking with 'arms-behind-the-back'. This is done with the knuckles of one hand clasped in the palm of the other and is a popular posture of high-status individuals, especially royalty and political leaders on formal occasions, when they are inspecting special displays prepared for them. The posture is one of extreme dominance because it is the opposite of the nervous 'body-cross' in which the arms are connected in some way across the front of the body as a kind of safety barrier. The hands-behind-the-back posture says that the person is so confident of his or her dominance that he or she has no need of even the slightest frontal protection. Schoolmasters also employ the same posture when walking in their school grounds, demonstrating their own dominance in that particular territory.

■ *Walking with the hands behind the back is a posture of dominance. It is the opposite of the nervous body-cross posture in which the hands are joined across the front of the body, creating a defensive barrier. In the hands-behind-the-back position the front of the body is completely exposed as the figure advances, revealing a sense of security and control.*

What the hands-behind-the-back posture implies.

An apparent flaw in this 'dominance' interpretation of the hands-behind-the-back posture arises from the custom of making humble private soldiers adopt this position when they are given the command to 'stand at ease'. But they are being forced to adopt a vulnerable position, which is very different from adopting one deliberately as a demonstration of fearlessness.

Other reasons for making contact with our own backs include the secretive use of hidden gestures, as when a child puts its hand behind its back in order to cross its fingers when telling a lie. The crossing of the fingers in these cases is an early Christian act of making a covert sign-of-the-cross to invoke God's protection when the lie is told.

We also do our best to scratch our own backs when they itch, but the difficulty in succeeding with this is summed up in the phrase 'you scratch my back and I'll scratch yours'. There is, indeed, very little back-contact we can administer on our own account, but there are a number of important ways in which others make contact for us, the most common of which is the proverbial 'pat on the back'.

A pat on the back is an almost universal way of demonstrating comfort, friendliness, congratulation or simple good humour. The reason why this act is so widespread, and always has the same meaning, is that it is a miniature version of that most basic of all person-to-person contacts, the embrace. When we are very young we enjoy the embrace of our mother's enfolding arms, spelling total security and love, and the feel of gentle hands pressing against our backs becomes one of the primary body signals of caring and friendship. We still indulge in full-blown body embracing and hugging when we are adults, if the situation is sufficiently intense and emotional, but in less passionate moments we switch to a minor version – the pat on the back – which reminds our bodies sufficiently of the major gesture. Even a brief, gentle pat given to someone in distress acts as a powerful comforting device, out of all proportion to the simplicity and brevity of the physical contact, because of the ancient echo it sounds from our infancy.

Another common form of contact is the back-guide, in which we steer a companion by lightly touching the back instead of the more usual forearm or elbow. The back-guide is slightly more intimate because it brings the two bodies closer together as they move forward. A related form of back-touching which does not involve actual steering is the light hand contact which 'lets you know I am here' when two people are standing together, facing in the same direction. The more intimate version of this is the back-rest in which the companion uses the back in front as a support.

Because of its large, flat expanse, the back is one of the most popular areas for detailed decoration such as tattooing. Magnificent demonstrations of the tattooist's art can be seen pricked painfully into the backs of brave men and women all around the world. The Japanese gangsters – the Yakuza – have the most beautiful examples, covering the entire back with swirling, rhythmic images, but the traditional designs of the European sailor also possess great charm. There is also a popular joke-motif tattoo which shows a hunting scene with horses and hounds chasing the fox down the length of the tattooed back, with the fox's tail just about to disappear between the buttocks.

THE BELLY

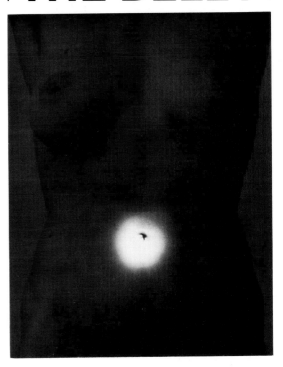

■ The human belly was accurately described by Dr Johnson as that part of the body which reaches from the breast to the thighs, containing the bowels. Although it excludes the genitals, it comes so close to the reproductive zone that it has been subjected to some degree of censorship. In Victorian times it became impolite even to use the word belly and a substitute had to be found. Because the belly region contains the stomach, and because the stomach is positioned high up in it, well away from the 'unspeakable' genitals, the Victorians decreed that a belly-ache should become a stomach-ache. A punch in the belly became a punch in the stomach. This anatomical inaccuracy became so entrenched that it has survived into modern times, even though Victorian prudery has long since been put behind us. In Victorian nurseries, even the word 'stomach' was considered too anatomical and was modified to 'tummy'. Stomach-ache became tummy-ache in the 1860s and that term too has stubbornly survived, reminding us that the Victorian heritage still lurks in the background of our supposedly liberated society.

While one class was shifting the belly politely up to the region of the stomach, another group was rudely pulling it down to the genitals. With equal and opposite inaccuracy they spoke of the belly as though it referred to the region below the pubic hair, rather than above it. An early slang expression for a mistress was a 'belly-piece' and the male's penis was called a 'belly-ruffian'. An 'itch in the belly' was sexual desire; 'belly-work' was copulation.

A third inaccuracy was to use the term belly as synonymous with womb. In the days when female criminals were executed for certain crimes there was a well-known strategy called the 'belly-plea' based on the ruling that a pregnant woman was spared capital punishment. In most prisons there were men called 'child-getters' whose less than

▼ In earlier days the male's pot-belly was proudly worn as a badge of success and well-being, indicating that he could afford to gorge himself on huge meals. Today, in advanced countries where the fear of starvation has been banished, the pot-belly has lost its grandeur and become a symbol of unhealthy self-indulgence.

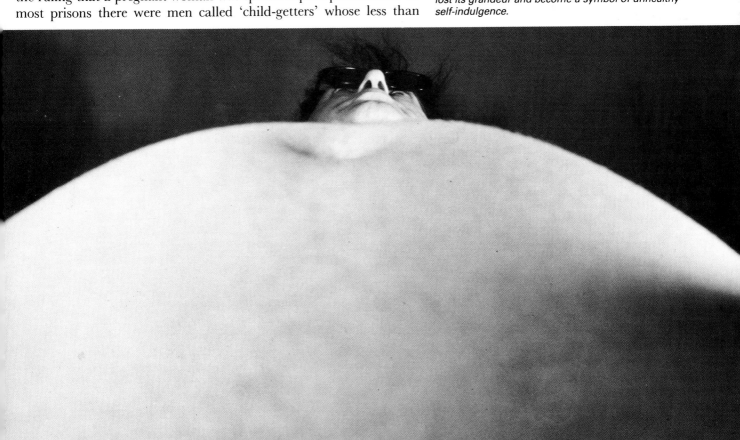

◄ Sumo wrestlers in Japan cultivate a gigantic belly for two reasons. It makes them heavier and therefore more able to heave their opponents out of the ring, and it also lowers their centre of gravity, which renders them more difficult to upend. They build up their great bellies by eating a mountain of special stew each day. The stew, called chanko-nabe, and made of fish, poultry, meat, eggs, vegetables, sugar and soy sauce is accompanied by 12 large bowls of rice and six pints of beer.

■ The big bellies of the West are less formally cultivated than those of the Orient. Professional wrestling, with its theatrical presentation, employs the massive paunch more as a psychological deterrent than a device for shifting the centre of gravity. Professional politicians whose bellies have expanded dramatically may also use them to add symbolic weight to their arguments.

arduous task was ensuring that inmates were qualified to 'plead their bellies'.

For our purposes we shall stick to the original Anglo-Saxon meaning of 'belly' – the lower front part of the body, below the chest and above the genitals, containing the stomach and intestines and, in the female, the uterus. In medical parlance: the abdomen.

This region of the body boasts few surface landmarks. Apart from the navel, or umbilicus (of which more later), there is the midline depression called the *linea alba*. In a typical adult this runs vertically from the navel up to the lower part of the chest. If a slender, athletic body is viewed in a suitable side-light, the *linea alba* shows up as a narrow but distinct indentation in the flesh, marking the place where the muscles of the left side of the body meet up with those of the right. In young muscular individuals the line can be detected below the navel as well as above it. However, anyone running to fat (at any age) will find it hard to detect either below or above the navel.

There are slight sex differences in the belly region. The belly of the female is more rounded in the lower part than that of the male. The female's belly is also proportionally longer than the male's, with a greater distance between the navel and the genitals. The typical female navel is also more deeply recessed than its male equivalent, assuming that both individuals are of similar average build. These differences can be summed up by saying that the human female has a

Two ways of cultivating the male belly.

larger and more curved abdomen than the male, a feature which is often exaggerated by artists. A young athletic male, by contrast, has a small flat inconspicuous belly which owes its sex appeal to its negative qualities.

As males and females grow older their bodies tend to become heavier and their bellies more generous. If they overindulge in what used to be called 'belly-cheer' or 'belly-timber' – in other words, food – they soon become ruefully or proudly big-bellied. In earlier days of shortage large bellies were often worn with pride and ostentation. Tribal girls were fattened up for their bridegrooms and successful industrialists would hang a gold watch-chain across their expanded waistcoats. The new body puritanism with its craving for eternal youth has changed all that. Now flat bellies are the order of the day for both sexes of all ages.

This change of belly-fashion has had one unusual side-effect. It

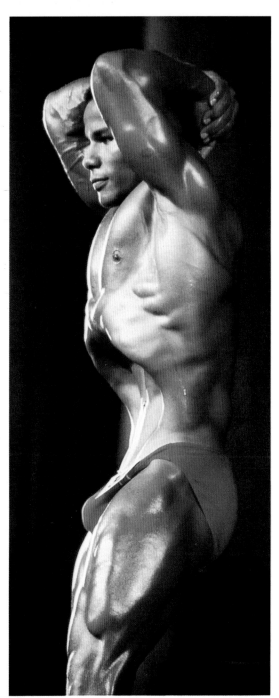

▲ *Belly-muscle control by body-builders has to be seen to be believed. Because the flat male belly has become the modern ideal for the exercise-puritan, health-fanatic, juvenile-worshipping urbanite of today, it follows that excessive flattening is called for where super-male displays are being given. But in many cases these experts do not know when to stop and their actions become so exaggerated that they start to lose their masculine appeal.*

The swelling contours of the maternal belly.

has altered the shape of the navel. On plump figures the navel is roughly circular in shape, but on slender ones it is more likely to be a vertical slit. A survey of works of art showing the more generously proportioned females of earlier days revealed that the vast majority (92 per cent) displayed circular navels. A similar survey of modern photographic models sees this figure fall to 54 per cent. So today's slender females are six times as likely to have vertical-slit navels as their more voluptuous predecessors.

There is more to this navel manoeuvre than mere loss of weight, however. All that the slimmer body does is to make the vertical navel a possibility. Whether it is displayed or not ultimately depends on the posture of the model. Even the skinniest female can present a circular navel if she slumps her body forward. So modern poses, either consciously or unconsciously, seem to be aiming towards a greater emphasis on the vertical-slit navel. The reason is not hard to discover. Because it looks like a body orifice, the navel has always played a minor role as a genital echo. Its indented shape in the centre of the belly makes it strongly reminiscent of the true orifices situated below it. The female's genital orifice is part of a vertical cleft, while her anal orifice is much more circular in shape. It follows that a shift towards a vertical navel display strengthens the specifically genital symbolism. In glamour photographs where the true genital cleft is concealed, the photographer and his model can contrive to offer a subliminal pseudo-orifice as a substitute for the real thing.

If this sounds rather fanciful, it is only necessary to look back at what happened to the navel in the more puritanical phases of the twentieth century. In early photographs it was studiously painted out, the pictures being retouched to give the ludicrous impression that the belly was completely smooth. This was done because, it was said, the navel was far too suggestive. Suggestive of what, was never mentioned.

With early films there was also shock and horror at the exposure of this part of the anatomy by dancing girls. An official letter from the censor to the makers of *The Arabian Nights* stated: 'Passed for adult distribution provided that all dancing scenes showing the dancing-girl's navel are cut out.' A second wave of film censorship, in the thirties and forties, returned to this suppression of the navel. The notorious Hollywood Code insisted that naked navels were to be outlawed. If they could not be covered in clothing, then they must be filled with jewels or some other kind of exotic adornment. What seemed to outrage the puritans about films in particular was that the dancers were able to *move* their navels – to make them gape and stretch as they undulated their half-clad bodies. This took the orifice symbolism too far and had to be ruthlessly stamped out to prevent sexual hysteria in the auditorium.

No sooner had the western world relaxed its censorship of the cinema navel than another assault was mounted. This time it occurred in the true homeland of the belly dance – the Middle East. There, with new cultural and religious moods sweeping the Arab world, night-club performers were instructed to cover their bellies when engaging in what had now come to be called 'traditional folklore dances.'

It is clear from these restrictions that the navel does have the power

■ The human female is less likely than the male to develop a pot-belly. When additional fat layers are added they tend to exaggerate the hip region, producing a wide belly rather than a prominent one. Even when the female belly does protrude, during pregnancy, its contour is subtly different from that of the overweight male.

■ When slender males and females are compared, the female belly region is seen to be longer and more rounded than that of the male. The distance between the female navel and the genitals is greater. This shapeliness of the female abdomen has made it a favourite subject for artists and a zone of major sexual attraction for the male. Its erotic impact is enhanced by its proximity to the genital region and by the presence of the orifice-like navel.

to transmit erotic signals, even if to most of us today it appears to be a comparatively innocuous detail of the human anatomy. Sex manuals have noted its appeal, stressing its fascination to young lovers who are exploring each other's bodies. The comments in such books reinforce its role as a genital echo. For instance: 'It...has a lot of cultivable sexual sensation; it fits the finger, glans or big toe, and merits careful attention when you kiss or touch.' (From *The Joy of Sex*.) A popular pose in illustrated sex manuals shows a man probing his partner's navel with his tongue – a pseudo-penis inserted into a pseudo-vagina.

Outside the sexual sphere the navel has caused something of a problem in religious circles. For those who believe in the literal truth of ancient religious texts there is the thorny problem of whether or not the very first man possessed a navel. If Adam (or his equivalent in other religions) was created by the deity, rather than born of woman, he presumably had no umbilical cord and therefore no navel. Early artists had the dilemma of deciding whether to include a navel on their paintings of Adam's body. Most of them opted to do so and no

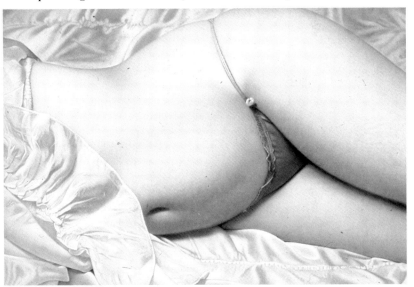

doubt invented their own reasons for the existence of Adam's navel. but their decision led to an even greater problem. For, since God created man in his own image, God too must have a navel; and naturally this gave rise to the further intriguing question: Who gave birth to God?

The Turks have found their own unusual solution to the problem of the first navel. They have an ancient legend which explains that, after Allah created the first man, the Devil was so angry that he spat on the body of the new arrival. His spittle landed on the centre of the man's belly and Allah reacted quickly by snatching out the polluted spot, to prevent the contamination spreading. His considerate action left a small hole where the spittle had been and that hole was the first navel.

A totally different symbolism sees the navel as the centre of the universe and it is in this loftier role that it is contemplated by Buddhist ascetics. To 'contemplate one's navel' has often been misinterpreted as a self-centred, inward-looking form of meditation, when in fact it

The more elongated the female navel the better it functions as a genital echo.

◄ *The shape of the female navel has changed in recent years, becoming more and more vertical. This is partly due to the more slender figures now in vogue and partly to the postures adopted. The result is a slit-shaped navel which strongly echoes the female genital cleft and therefore transmits a more potent sexual signal.*

▼ *Belly-dancing dates back to the time when harem girls straddled the bloated bodies of their masters and made vigorous pelvic movements to aid the sexual consummation of their sexual encounters. Today the undulating, grinding and thrusting actions of the dancer's abdomen present a formalized version of the early harem copulation movements, although the actions are usually referred to merely as 'traditional folk-dance movements'. This dancer's costume draws attention to the navel and gives it special significance by alternately concealing and revealing it.*

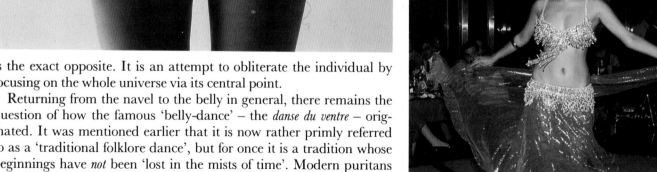

is the exact opposite. It is an attempt to obliterate the individual by focusing on the whole universe via its central point.

Returning from the navel to the belly in general, there remains the question of how the famous 'belly-dance' – the *danse du ventre* – originated. It was mentioned earlier that it is now rather primly referred to as a 'traditional folklore dance', but for once it is a tradition whose beginnings have *not* been 'lost in the mists of time'. Modern puritans might prefer in this case that they *had* been lost.

There are three main movements in the belly-dance: bumps, grinds and ripples. Bumps are forward jerks of the pelvis. Grinds are rotations of the pelvis. And ripples are muscular undulations of the belly region involving expert muscular control. The first two are easy and commonplace. The third is the province of only the most skilled performers. All three are active sexual movements. They began in the harem, where the overlord was usually grossly fat, hopelessly unathletic and sexually bored. To stimulate him sexually, his young females would have to squat over his recumbent body, insert his penis

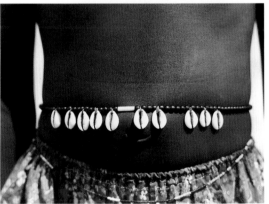

■ The belly region is not one of the most favoured zones for body tattooing because it is usually less 'exposable' than the chest, arms or back, and is therefore difficult to display to admiring companions. But when a tattooee goes to the extreme of full-body decoration, the belly zone is capable of providing a splendid canvas for the tattooist's art. In parts of the world where the belly is normally exposed, however, its decoration is widespread and may often, as the shells do here, reinforce the genital echo provided by the navel.

and then wriggle their bodies enticingly to bring him to a climax. This wriggling became an expert activity, with special movements of the female pelvis and contractions of the abdominal muscles to massage the great lord's penis. As an act of copulation it has been described as 'fertile masturbation'.

As time passed, the female pelvic movements became developed into a visual display to titillate and excite the master of the harem before copulation itself was attempted. Freed from contact with his sluggish body, the harem girls were able to exaggerate the actions and make them more rhythmic. With music added the whole display soon became stylized into what was called the 'muscle dance', and what we today refer to as belly-dancing.

Some sources have suggested an additional element. They claim that certain of the movements represent not copulation but birth. It is pointed out that in many cultures, before a pregnant woman was turned into a doctor's patient, she did not lie down to give birth, painfully pushing against gravity, but instead adopted a squatting position using gravity to help her deliver her child. She assisted the birth by moving her abdomen in a rolling motion, bearing down hard as she did so. It is this element of parturition that is said to have been incorporated in the belly-dance as the centuries passed. It became not merely a dance of mimed copulation by a vigorous young female straddling an indolent, corpulent male, but a symbolic enactment of both conception and birth – the whole reproductive cycle in one performance.

Whether this modified interpretation of the belly-dance is true, or whether it is an attempt to sanitize a purely copulatory dance and bring it into line with other 'folkloric' activities, is hard to say. In any case, the purification process has gone much further in recent years. A belly-dancing instruction manual published in the 1980s introduces its subject with the following words: 'In its new role as a healthy physical art form, the emphasis is on its keep fit qualities.' The harem girl has become a gym mistress.

Despite the fact that the belly dance is now being promoted as 'an excellent form of therapy for tension and depression', the names that have been assigned to its various movements still give a vivid glimpse of its more erotic origins. They include: the pivoting hip rotation, the travelling pelvic roll, the undulating pelvic tilt, the heel-hip thrust, the backbend shimmy, the hip skip and the camel rock. Clearly, all has not been lost.

In its non-sexual symbolism, the belly, like the navel, has several roles. Its most widespread association is with the earthier, animal side of man. Because the belly is, through its contents, connected with our appetite for food, it becomes linked with all our animal appetites. There is a Greek proverb which states: 'The vilest of beasts is the belly.' And another comment from ancient Greece cries out: 'May God look with hatred on the belly and its food; it is through them that chastity breaks down.' And Nietzsche added: 'The belly is the reason why man does not mistake himself for a God.'

This unflattering western symbolism is at complete odds with oriental symbolism which sees the belly as the seat of life. In Japan the belly is regarded as the centre of the body and it is for this reason that ritual suicide there is directed towards it. The Japanese act of hara-

Cheek-to-cheek dancing anticipates the belly contacts of tender love-making.

▶ The major form of belly contact between adults occurs during frontal copulation, although this is anticipated in a mild version when couples dance cheek-to-cheek. It was because of its belly intimacy that the waltz was so savagely attacked when it first appeared. 'Pollution has entered the ballroom,' was the cry. 'The proximity of the partners makes it graceless and vulgar...A fine exhibition truly for the zoological gardens...but not for the select ballroom.'

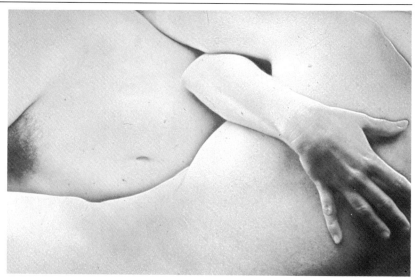

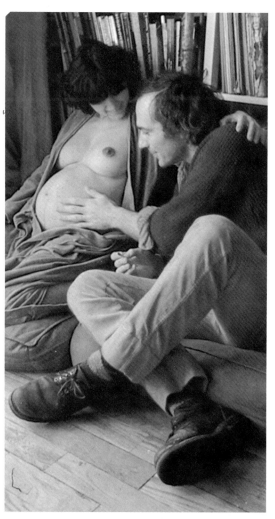

▲ The most tender form of belly contact occurs when a husband wonderingly places his hand on the hugely swollen belly of his heavily pregnant wife. It is often this gentle touch which first triggers off the full emotional pride of impending paternity, and converts the clumsiness of a thick-waisted wife into the serene rotundity of a mother goddess.

kiri consists of self-disembowelling with a sharp sword. This is so agonizingly inefficient that an assistant has to stand by to decapitate the suicide figure, as a way of swiftly putting him out of his misery. If it were not for the fact that the belly is the all-important seat of life, a more efficient method of self-despatch could easily have been found. (A literal translation of hara-kiri is belly-cut).

Ordinary, everyday belly gestures are few and far between. We occasionally clasp or embrace our bellies when we feel slightly threatened by our companions. The arms act as a form of barrier signal across the front of our bodies. Unconsciously they say: 'I must protect my soft underbelly from possible attack.' This is a variant form of the more typical body-cross which sees the arms folded firmly across the chest. All such body barrier movements indicate an uncomfortable social mood, where slight unease is present in personal relationships. Having said this, however, it is important to establish that the person performing the self-hug of his belly has not just eaten a large number of unripe apples, in which case the interpretation of the gesture may have more to do with belly-ache than subliminal belly protection.

Because of its proximity to the genital region, the belly figures rarely in interpersonal contacts. When someone touches someone else's belly, the two involved are usually members of same family, lovers, or very old friends. Parents may pat their children's bellies when they have eaten well; a proud husband may gently pat the protruding belly of his pregnant wife to show his pleasure at her condition; two old drinking friends may pat each other's beer bellies as comments on their ever-increasing girth; and lovers may lie quietly together with the head of one resting on the recumbent belly of the other. Apart from these actions and a rare punch in the belly from an enemy, there is only one other important person-to-person contact in this category and that is the belly-to-belly intimacy of frontal copulation. Strangely enough this posture is the subject of one of the oldest recorded jokes known to mankind. In one of the ancient Sumerian texts dating from the third millennium BC the writer records with sad humour: 'Brick on brick was this house built; belly on belly was it torn asunder.'

THE HIPS

■ The broad human pelvis produces a widening of the body at the point where the trunk meets the legs. We call this projection in the body contour the hip, deriving the name from the verb 'to hop'. Because the female pelvic girdle is wider than that of the male, women have wider hips, and this basic biological gender difference has led to a great variety of exaggerations and modifications.

If a female wishes to intensify this type of feminine body signal she can do so directly by padding out the hips and making them even wider, or she can do so indirectly by tightening the waist. In both cases the projection of the hips is enhanced and it appears to be this degree of projection rather than the absolute size of the hips which tranmits the primeval gender signal to the male.

Today most women are prepared to rely on natural unmagnified hip signals, but in the past they have often made themselves the slaves of super-hips, and the victims of super-hip technology. The lengths to which some fanatics went is hard to believe. In the sixteenth century, European dress stores were busy selling huge cumbersome 'hip cushions' looking like padded motor-tyres and almost as big. These were tied on beneath the vast skirts to double the natural pelvic width. They made any kind of vigorous or athletic movement impossible and created a costume so heavy and tiring that the ladies of the day were incapable of any energetic activities.

Worse was to come. In the seventeenth century the custom of tight-lacing reached such a pitch that many a young virgin of high status would not be satisfied until she could encircle her own waist with her hands. Writing in 1654, Thomas Bulwer was horrified at the damage caused by the practice, complaining that when they 'shut up their waists in a Whale-bone prison...they open the door to Consumptions, and a withering rottenness.' His words were ignored, and during the next century we see the steady rise in importance and wealth of the staymaker, who came to rival in social standing the dressmaker and the coiffeur. His professional task of encasing the ladies of quality in their breathtakingly-tight body-prisons gave him a degree of intimacy that even the ladies' beaux were not allowed to enjoy.

At the end of the eighteenth century the craze for tight-lacing receded and young women could breathe again. But not for long. Within a few decades heavy corseting was back and remained in force throughout much of the Victorian period.

In late Victorian days, a beautiful girl was one whose waist measurement in inches was the same as her age at her last birthday. To achieve this difficult goal many young ladies of fashion had to wear their tightly-laced corsets twenty-four hours a day, though how they managed to sleep in them is a mystery. In extreme cases, certain women even had their lower ribs surgically removed to help reduce their waist circumference. (The smallest recorded waist measurement for any adult female of normal height is 13 inches.)

Critics of the corset became even more vociferous than they had been before, listing all the kinds of damage done by the 'mutilation' of tight bindings. The corsets, it was claimed, sapped vitality, clawed at the tender flesh, displaced inner organs, deformed the rib-cage and restricted breathing. Inevitably they also caused miscarriages. Some doctors demanded reform but many others, although unhappy,

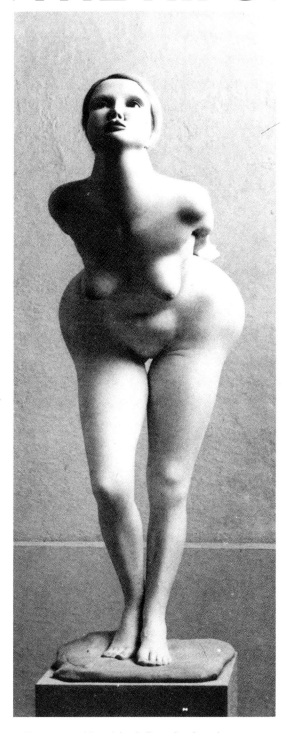

■ Because a wide pelvic girdle makes it easier to give birth, broad hips have become symbols of female fertility. They have also become powerful signals of sexual attraction and have often been artificially exaggerated by special padding and other costume devices. Sculptors can increase the femininity of their nude figures by enlarging the hip zone slightly beyond the natural level, although there is a danger that this may become too extreme to be erotically appealing.

Hourglass figures embody a 'bondage' factor.

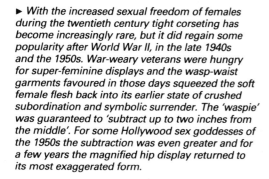

▲ *The hourglass figures of the turn of the century were popular not only because they magnified the feminine hip-to-waist ratio, but because they also embodied an unspoken 'bondage' factor. The corsets 'trapped' the soft flesh of the attractive young female, giving her the air of someone ensnared and vulnerable, unable to move freely or forcefully.*

▶ *With the increased sexual freedom of females during the twentieth century tight corseting has become increasingly rare, but it did regain some popularity after World War II, in the late 1940s and the 1950s. War-weary veterans were hungry for super-feminine displays and the wasp-waist garments favoured in those days squeezed the soft female flesh back into its earlier state of crushed subordination and symbolic surrender. The 'waspie' was guaranteed to 'subtract up to two inches from the middle'. For some Hollywood sex goddesses of the 1950s the subtraction was even greater and for a few years the magnified hip display returned to its most exaggerated form.*

were prepared to overlook a few physical sacrifices in the service of high fashion. Some to their shame actually cashed in on the craze and sold special 'health corsets'. The most bizarre were Dr Scott's Electric Corsets of 1883. These were said to be 'constructed on scientific principles, generating an exhilarating, health-giving current to the whole system. Their therapeutic value is unquestioned, and they quickly cure, in a marvellous manner, Nervous Debility, Spinal Complaints, Rheumatism, Paralysis, Numbness, Dyspepsia, Liver and Kidney Troubles, Impaired Circulation, Constipation and all other diseases peculiar to women'. The electromagnetic force which was supposed to work such wonders was vouched for by no less a person than the Surgeon-General of the United States. Yet the sad truth was that Dr Scott's corsets caused many of the ailments they were said to cure.

By the end of the nineteenth century further advances had been made in corset technology. In 1901 an advertisement showed a charming young lady swimming in the sea beneath a headline which read: Warner's Rust-proof Summer Corsets.

Slimline hips suggest youthful innocence.

A curious feature of the long period of fanatical tight-lacing and tight-hooking was the way in which it lost sight of its original purpose, which was to render the female more feminine and therefore more sexually appealing. Victorian females *without* stiff corseting were looked upon as wanton and depraved, despite the fact that it was they who were displaying themselves in their natural shapes. Their waists and hips were not being exploited as super-gender-signals, but this was completely overlooked. The reason is not hard to find. Tight corsets had become so much a part of polite fashion that to appear without them was tantamount to having started to undress in public.

Perhaps, in the end, it was this idea of equating sexual freedom with freedom from corsets that was their undoing. They gradually disappeared during the twentieth century, re-surfacing only briefly in highly simplified form during such fashion phases as the 'New Look'

■ In recent years there has been a strong shift towards a mock-juvenile contour, with adult women offering the slim lines of young schoolgirls as a sexual attraction. By omitting the broad childbearing hips this display suggests athletic fun and sex rather than the heavy responsibility of breeding and the establishment of a family; and by imitating the weaker 'little girl' frame it also signals the subservient immaturity of the juvenile. For some female athletes the hip-slimming process goes so far that they begin to transmit pseudomasculine signals. Their broad shoulders and narrow hips may indicate supreme health and fitness, but their body contours lose their erotic impact.

Ludicrous extremes of waist-tightening.

A CORRECT VIEW OF THE NEW MACHINE FOR WINDING UP THE LADIES

in the late 1940s. The era of female torso strangulation was over.

If we scoff at these earlier fanatical attempts to create a slender fe-male waist, it is as well to remember that we have our own masochistic equivalent today. Modern females may have discarded the crippling clothing restrictions of previous generations, but they have replaced the old tyranny with a new one: the rigid diet. By keeping unnaturally slim they manage to reduce waist measurement by cannibalizing their natural fat layers.

As with Victorian corsets, the official excuse given for modern crash diets is 'improved health and vigour'. The hidden truth, as before, is the primeval sexuality of the small female waist. If a teenage female has a 22 inch waist before she becomes a mother, then after two or three children she can expect it to widen out to between 28 and 30 inches, unless she takes special steps to prevent this natural in-crease. It is because of this change in girth that very narrow waists have always symbolized virginity and innocent youth. It is important to realize that a narrow waist not only signals 'I am female', but it also quite specifically says 'and I am not pregnant'. To put it another way, the narrow waist proclaims 'I am capable of being made pregnant', which in some contexts becomes 'You may make me pregnant.' The bulky torso of an overweight woman is too reminiscent of the bulging shape of a mother-to-be to be able to transmit clear-cut sig-nals of sexual invitation to prospective mates. It is because of this that the slender-waisted bodyshape has such a powerful impact on the human male.

Despite this it is important not to go too far in the quest for skinny elegance. The top model girl aptly named Twiggy had the vital stat-istics 30½-24-33 (chest-waist-hips, in inches). The average British female during the same period (in the 1970s) had measurements reading 37-27¾-39. Twiggy scored sexuality-points because her waist was narrower, but at the same time she lost points because there was less contrast between her waist and her hips. She sent out enticing signals of slender-waisted virginity, but the average female sent out equally appealing signals of curvaceous nubility.

The trick, of course, is to send out both at once, but that is not easy.

When in tomboy mood, female fashion tries to abolish the waist-and-hips contour.

■ *During the last century the tightening of the waist to exaggerate the hip-line reached ludicrous extremes. Throughout much of the Victorian epoch young women suffered agonies in their quest for the perfect shape, and it is hardly surprising that fainting and swooning away became so common-place in polite society. Although widely popular, these fashions were often severely criticized, the corseted females being referred to scathingly as 'ants' or 'bottle-spiders'.*

▲ *With the increasing freedom of the twentieth century, the earlier narrow waists and padded hips gave way to more shapeless flapper costumes in the 1920s. In that playful, tomboy decade, the females abolished the waist-and-hip contour and displayed themselves more as naughty little children, offering their bodies for fun and games rather than serious maternity. If hips were indicated on their costumes, they were usually displaced, as if suggesting that they were no more than a token gesture to feminine shape.*

Only the most perfectly balanced females, those who are as slender as is possible without sacrificing feminine curves, will be able to score on both counts. Such individuals do exist, even without the benefit of Victorian corsets. At the time when Twiggy was epitomizing the emaciated fashion model, Miss World contestants were parading with figures that gave average measurements of 36-24-36. Their waists were as slender as Twiggy's, but their hips were much wider, and the difference between their waist and hip measurements was greater than either Twiggy's or the average female-in-the-street. Purely at the level of body-contour signalling they had a clear advantage.

Turning from shapes to movements and postures, it is not surprising to find that most hip actions have a strong female bias. Walking styles which involve conspicuous hip sway or wiggle are so strongly feminine that they are employed as caricature elements in comically erotic theatrical performances. Everyone from Mae West to Marilyn Monroe has swayed a sexy hip from time to time. But only males impersonating females, or outrageously camp homosexuals would allow masculine hips to undulate in this way.

The sideways hip-jut is equally feminine, or effeminate. This is the postural equivalent of the walking hip-sway. It is a slightly contradictory pose, being both contrived and relaxed. The message it transmits is 'Look what nice hips I've got', but the off-balance, asymmetrical posture of the body fails to signal a clear-cut mood.

Many dance routines incorporate vigorous hip movements, such as the side-jerk, the rapid hip-shake and the hip rotation, and again these are usually the province of the female rather than the male. In the famous Hawaiian Hula Dance young girls, their pelvic regions emphasized by grass skirts, perform a variety of rhythmic movements of the hips, swaying them, jerking them and rotating them in time to the music. Two special dance movements are the *Ami* and the *Around-the-island*. The Ami is a Hip Roll. One hand is raised while the other rests on the hip. The hips are then moved in a circle, first clockwise and then counter-clockwise. The movement called Around-the-island is similar except that the body turns a quarter-circle with each hip roll, going 'around the island' in four counts.

Typically, male dancers avoid such movements, but there have been exceptions to this rule. When Elvis Presley first began to perform in public he created a scandal because of his hip gyrations. His use of hip rolls and jerks was so exaggerated he was quickly dubbed Elvis the Pelvis and in his early TV appearances his hips were banned from the screen. He was too aggressively masculine for his actions to have been offensive as pseudo-female. Instead they were interpreted as mimics of copulatory pelvic thrusting, and it was some years before he was permitted to display his full routine on public screens. What Presley had proved, almost single-handed, was that rhythmic hip movements are by no means the exclusive province of females. As already mentioned, they had become strongly associated with female dancing because there is more natural sway to female hips than male ones. When this sway was exaggerated and stylized in such dances as the Hula, the male copulatory role of hip jerks and sways was overlooked. Presley corrected this and caused a sensation in the process. It was his good fortune that the feminine label attached to

Hip movements and pelvic flexibility.

hip movements had kept any other males from risking them in a stage performance.

Being a hippie, it should be mentioned, has nothing to do with the hips of the body. The long-haired, drug-taking, love-promoting hippies of the 1960s derived their name from the word 'hipster'. A hipster was a 1950s jazz and swing term for someone who was 'in the know', or 'hip'. The word 'hip' in this context comes from turn-of-the-century military slang. When the drill sergeant called out 'hip-two-three-four', the group that were well synchronized were said to be 'hip'. They knew what they were doing because they all brought their legs 'hip' (=up) together. The later breed of hippies would have been horrified to have known that their anti-military movement had as its title a term relating to military efficiency.

Among the regional hip gestures, of which there are very few, there is a strongly copulatory one, found as far apart as South America and the Middle East. In it, the elbows are held tight against the hips with the forearms pointing forward. The fists are clenched. The hips then jerk forward. In the South American version the arms are jerked back as the hips move forward. In the Middle East version the arms stay still while the hips jerk rapidly back and forth. It looks as though the two gestures, both of which signal 'copulation', developed independently of each other and their similarity is due to the fact that they are both simple mimics of the act of male pelvic thrusting.

Perhaps the most important hip gesture of all is the hands-on-hips action called the akimbo posture. It is usually said to indicate authority, defiance or an emphatic mood, but it is more complex than that. Essentially it is an anti-social gesture. It is the opposite of holding out the arms to invite an embrace. It is, indeed, quite difficult to embrace someone who stubbornly retains an akimbo stance. When the hands are placed firmly on the projecting hips, the elbows stick out sideways like arrowheads pointing away from the torso. It is as if they are saying, 'Keep away, keep back, or I will jab you.' It is performed unthinkingly and automatically when the mood is right, with the performer hardly aware that he or she has done it. And it is global in distribution.

The akimbo action occurs whenever the person concerned is in a mood of rejection. This is why it is characteristic of defiance. The guard stands at the door, arms akimbo, silently saying 'Keep away, no one shall dare to come near to me.' This is also why it carries a mood of authority. The person who holds authority and wishes to display it must be seen to be standing apart from others, not sharing space and posture with them. The akimbo gesture of a dominant member of a group tells the others to keep their places.

The akimbo gesture is also used by individuals who have just suffered a set-back. They may not be in a dominant mood but they are certainly not seeking fellow-comfort. The footballer who has just lost a goal immediately adopts the hand-on-hips posture, usually with his head slightly lowered reflecting his lowered spirits. His akimbo message is 'Keep away from me, I am so angry I don't want anyone near me.'

If a person wishes to dissociate himself or herself from a group of people on, say, his or her left, then only the left arm comes up into the akimbo posture. If there is a group on the right with which the person feels some affinity, then the arm on that side will stay down. These

■ *An alternative to the hourglass display is the dynamic hip jut and sway of modern dancing. Here the proportions of the hips in relation to the waist need not present such a powerful contrast, attention to the hips being attracted more by their amplified movements. Because dances involving hip swaying have traditionally been female rather than male, exploiting the amazing flexibility of the female hip joints, it came as something of a shock when rock-'n'-roll performers began to gyrate their pelvic girdles while singing. They avoided accusations of effeminacy, however, for the simple reason that their hip-jerks carried strong echoes of masculine pelvic thrusting during copulation.*

The anti-social hands-on-hips posture.

■ *The akimbo posture of hand-on-hip is an unconscious body signal used all over the world. It is an anti-social sign, saying 'Keep away'. If performed by a person standing alone it sends out a rejection message to all-comers. If it is performed by someone deep in close conversation it acts as a signal telling us we are excluded from that conversation. The elbow projects from the side of a body like an arrowhead pointing in the direction we should go.*

half-akimbos, often seen at parties or other social gatherings, quickly reveal the ties which certain individuals feel with the others present.

A strange feature of the akimbo posture is that, despite its world-wide use, it does not seem to have a name in other languages. It is given a descriptive phrase such as 'fists-on-haunches' or 'the pot with two handles', but there is no single-word equivalent. This reveals the extent to which the posture is taken for granted. It is one of those common human behaviour patterns which we see every day and to which we react subliminally without ever analysing the body signal we are receiving. If it was a more conscious gesture like a salute or a wave, there would be words for it in every language.

Finally, there is the inter-personal contact of the hip embrace or waist embrace. Young lovers often walk for long distances together, progressing snugly side by side, with their flanks touching and with their arms around each other's waists. Their embracing hands come to rest on their partner's hips. It is as if they are attempting to indulge in a full embrace and walk along at the same time, with the hip embrace as the compromise solution. It is not an easy double action to perform and hampers forward movement, but on such occasions the mobility of the couple is less important than their display of intimacy – which is done both for themselves and for the others around them. It acts as a powerful 'excluder' to anyone accompanying them or watching them.

As a tie-sign, this kind of embracing has a more potent message than the equally common shoulder embrace. Two males may indulge in a shoulder embrace, either standing together or walking along. It is friendly, but there is nothing so intimate about it that it suggests a sexual liaison. To embrace the waist of another person, however, brings the embracing hand much nearer to the primary sexual zone and makes the action more sexually loaded. For this reason males offer such an embrace only to females, unless of course they are intent on displaying their homosexuality in public.

A sample of waist and hip embraces was analysed for gender differences and it was found that in the majority of cases only one partner was actively embracing at any one time. The other was permitting the embrace but was not reciprocating. In 77 per cent of cases the embrace was by a male to a female; in 14 per cent it was by a female to a male; and in 9 per cent it was by a female to another female. (Parents embracing their small children were excluded from this sample.) As anticipated, there were no males embracing males, but it seems that the taboo on same-sex waist-embracing is less strict between females than between males. In this respect, however, it does not differ from many other public intimacies, such as greeting kisses.

The big difference in percentages between male-to-female and female-to-male embraces neatly sums up the whole attitude of adults to the waist and hip region of the body. The males are obviously much more interested in embracing the female waist than vice versa. Viewed in social terms it is now clear that the waist and hips are essentially female attributes. Because of the broad child-bearing pelvis which makes them such a strong feature of the female form they are almost as loaded with femininity as the breasts. Only when the male starts to make his urgent pelvic thrusts do they begin to acquire a more masculine flavour.

THE BUTTOCKS

■ The buttocks have quite unfairly become the 'joke' region of the human body. They make people laugh; they are a popular subject for dirty jokes. The behind, the backside, the bum, the buns, the arse, the rump, the bottom – whatever name they are given, the buttocks are looked upon as either ridiculous or obscene. Even when they are considered as an erotic zone, because of their proximity to the genitals, they are more likely to be pinched or slapped than caressed.

This negative attitude persists despite the fact that the buttocks are specially and uniquely human. We acquired them when we took that truly giant step for mankind – and stood up on our hind legs. The powerful, bulging gluteal muscles expanded dramatically, enabling our bodies to remain permanently and fully erect, and it is these muscles which give us the pair of curved hemispheres at the base of the back which we ungratefully find so laughable.

It is easy to see how this has come about. The buttocks are not alone. Between them lurks the anus, through which must pass, day after day, all our solid waste matter and – even more notoriously – the occasional emission of gas. Furthermore, when we bend down, the genitals swing into view, also framed by the twin curves of the buttocks. So there is no escaping excretory and sexual associations.

It follows from this that to display the buttocks is interpreted either as a gross insult – a symbolic act of defecation on an enemy – or as a gross obscenity – a shameless presentation of sexual organs. In modern society the showing of a bare behind in public can produce reactions varying from embarrassed laughter to serious complaints, outrage and even prosecution. Quite recently in Switzerland the Federal Supreme Court was wrestling with the fine point as to whether a particular buttock display was 'offensive' or 'indecent'. On this subtle distinction rested the decision concerning a conviction. A Swiss woman, during an angry quarrel with a neighbour, had suddenly 'displayed her naked posterior'. As there were children present she was arrested on a charge of public indecency and had been found guilty by a lower court. After due deliberation the Supreme Court quashed the woman's conviction and even awarded her costs. They did this because they came to the conclusion that 'the gesture was

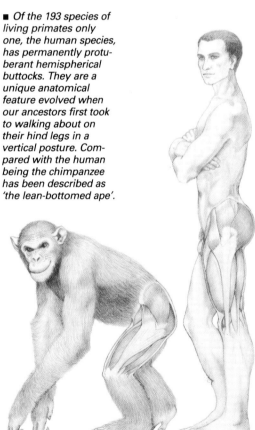

■ Of the 193 species of living primates only one, the human species, has permanently protuberant hemispherical buttocks. They are a unique anatomical feature evolved when our ancestors first took to walking about on their hind legs in a vertical posture. Compared with the human being the chimpanzee has been described as 'the lean-bottomed ape'.

◄ The deliberate exposure of the buttocks as a joke insult is occasionally used by the more outrageous stars of the worlds of pop music, cinema and sport. Onlookers laughing at these displays rarely appreciate their historical origins.

Buttock beauty is a matter of deep-rooted significance to both males and females.

▼ The erotic beauty of the female buttocks was so important in classical times that the Greeks had a word for it: kallipygia. They even built a temple to the 'Goddess with Beautiful Buttocks', who was given the name Aphrodite Kallipygos, and a statue of her lifting her dress to display her charms (see inset) can be seen today in the National Museum at Naples. Ever since those days artists have enjoyed posing their models to emphasize this part of the anatomy. By carefully concealing any anal or genital details in the cleft between the buttocks, they have been able to deflect attention more towards the aesthetic curvature of the female pelvic region, reducing its sexuality to an implicit rather than an explicit role. But even with genital concealment, the shape of the buttocks alone carries a powerful erotic message, dating back to the time when the aroused male always mounted the female from the rear.

certainly insulting behaviour and punishable as such but it could not be considered indecent because no organs of procreation were involved'. Presumably, if she had bent further forward when making her defiant gesture, her conviction would have stood.

Such extreme responses to buttock displays are rare today in the West. Streakers who drop their trousers or lift their skirts at sporting events usually produce only laughter, as do 'mooners' at American colleges who stick their behinds out of dormitory windows. As a protest, nudity is not what it used to be.

The buttock display is sometimes made more abusive by the addition of the phrase 'kiss my arse'. Taken at face value this is insulting because it demands a humiliating act of subordination. But there is more to it than that. Although probably neither the insulter or the insulted person realize it, they are engaged in a modern version of an age-old occult practice. To understand this it is best to return first to ancient Greece.

The present-day view of the buttocks as a joke region of the body was not shared by early Greeks. To them it was an unusually beautiful part of the anatomy, partly because of its pleasing curvature but also because it made a powerful contrast with the animal rump of apes

In today's more honest sexual climate, females are now admitting that they have always found male buttocks an exciting erotic zone. Their preference is generally for tight, compact male buttocks, small but muscular. This response, revealed in replies to scientific questionnaires, embodies three elements. The females prefer smaller male buttocks because this offers a contrast with the wide female buttocks. Narrow male buttocks are therefore essentially masculine. Secondly, they like them to be strong and muscular because this reflects male power and vigour. And thirdly, they also like them to be muscular for a specific sexual reason, because muscular buttocks promise athletic pelvic thrusting during copulation.

■ *The phrase 'kiss my arse' has an ancient origin connected with devil worship. Today it is no more than an insulting demand for an act of humiliating servility, but in earlier times it referred to the way witches were believed to pay homage to the devil, who had a face on his buttockless rump, with a mouth in place of an anus.*

and monkeys. The human hemispheres were so different from the tough patches of hardened skin (the ischial callosities) on the lean-bottomed ape, that the Greeks saw them, quite correctly, as supremely human and non-bestial. The curvaceous Goddess of Love, Aphrodite Kallipygos – literally the 'Goddess with Beautiful Buttocks' – was said to have a behind more aesthetically pleasing than any other part of her anatomy. It was so revered that a temple was built in its honour – thereby making the buttocks the only part of the human body so honoured.

This early view of the buttocks as exquisitely human gave rise to a further notion. It was argued that if rounded buttocks were the hall-mark distinguishing mankind from the beasts, then the monsters of darkness must lack this particular anatomical feature. So it was that the devil gained the lasting reputation of being buttockless. Early Europeans were quite convinced that the devil, even though he could assume human form, could never complete the transformation because, try as he might, he could never manage to simulate the rounded human buttocks. This, the most gloriously, most exclusively human feature of the body, was beyond even his fiendish powers.

This weakness was thought to be a source of great anguish to the devil, and it provided a golden opportunity for tormenting him. To inflame his envy all that was needed was to show him your bare buttocks. Because it reminded him of his deficiency, the sudden display would force him to look away, averting his evil gaze. This protected the buttock-displaying human from the much feared 'Evil Eye' and became widely employed as a valuable device for repelling the forces of wickedness.

Buttock displays used in this special way were not regarded as vulgar or dissolute. Even Martin Luther employed this method of defence when he was tormented by nightly visions of the devil. Early fortifications and churches often displayed carvings of human figures showing off their rounded buttocks to drive away evil spirits, the exposed behinds always pointing outwards from the main entrances. In the Germany of those times, if there was a particularly terrifying storm at night, men and women alike would protrude their naked buttocks from their front doors in the hope of warding off the powers of evil and avoiding a stormy death.

It is quite probable that this is how all buttock displays began and that today's streakers and mooners are carrying on an ancient Christian tradition without realizing it. With the devil out of fashion as the great enemy, the display is now seen merely as 'rude'. From an act of religious defiance it has slipped easily into an obscene exposure of a taboo body zone.

But how does this explain the phrase 'kiss my arse'? To understand this it is necessary to examine early engravings depicting the devil. If he does not have buttocks, then what *does* he have on his hind quarters? The answer is that where his buttocks should be he has another face. This second face is the one which was supposed to be kissed by witches as part of the ritual of the Sabbath. Accused of the filthy action of kissing the devil's rump, they reputedly defended themselves by insisting that they had only kissed the mouth of his second face.

All these activities were, of course, the inventions of fertile medieval imaginations, but that is beside the point. The legends and beliefs

A permanent buttock-display is a distinctively human anatomical feature.

handed down from one superstitious generation to the next, made it clear that 'arse-kissing' was the foul act of a follower of Satan and, as such, was an abhorrence. When the superstitions began to fade and die away, the connexions were lost but, as so often happens, the popular phrase survived to be incorporated in the modern insult.

So far, the display of the buttocks has been examined purely as a hostile act – as ancient defiance or as modern insult. But there is another side to it. In a completely different context, the buttocks also transmit powerful signals of sexual appeal.

The females of many species of monkeys and apes have brightly coloured rumps. Their hind quarters become increasingly conspicuous and swollen as the time of ovulation approaches, then recede again as it passes. This means that a male can tell at a glance whether a female is sexually active. Matings usually take place only when the females are displaying their most exaggerated sexual swellings.

Human females are different. Their rumps do not rise and fall with their menstrual cycles. Their buttocks remain protuberant throughout. Matching this, sexuality also remains high. As part of her pair-bonding system, the human female has extended her sexiness so that she is always potentially responsive to the male. She will mate even at times when she cannot possibly conceive, because the function of human mating is no longer purely procreative. As a reward system it helps to cement the bond of attachment between male and female, keeping the vital family breeding unit together. For humans, copulation is *literally* love-making, and it is important for the female's body to be able to transmit its erotic signals at all times.

It could be argued that if the gluteal muscles of the human rump are essentially concerned with the mechanics of standing erect, females cannot help but display permanently protruding buttocks. In their sexuality, however, the female buttocks go beyond the demands of simple mechanics. Relative to body size they are larger than those of the male, not because they are more muscled but because they incorporate much more fatty tissue. This additional fat has been described as an emergency food store – rather like a camel's hump – but whether this is true or not, the fact that it is gender-linked automatically makes it a female sex signal. It is accentuated by two other female properties: the backward rotation the pelvis and the sway of the hips in walking. The typical female – not to be confused with the female athlete, whose body has been severely masculinized by special training – has a more arched back than the male. In a normal position of rest her behind sticks out backwards more than the male counterpart, regardless of its size. When she walks, the different leg and hip design of the female skeleton produces a greater undulation in the buttock region. To put it bluntly, she wiggles as she walks.

When these three qualities – more fat, more protrusion and more undulation – are combined, the result is a powerful erotic signal to the male. This is not because the female is deliberately thrusting out her behind and consciously wagging it at admiring males but simply because this is the way her body is designed. She can, of course, exaggerate her natural signals if she is prepared to risk caricature, and wiggle her bottom outrageously. But even if she does nothing at all her basic anatomy will be constantly transmitting signals on behalf of her gender.

■ *The hemispherical shape of the female human buttocks acts as a permanent sexual display. By contrast, the rump display of other female primates varies dramatically during the monthly cycle, transmitting sexual signals only during the brief period of ovulation. At this time, conspicuous sexual swellings are present, but then they subside for the rest of the cycle, when the female cannot be fertilized. The sexual 'swellings' of the human female – her rounded buttocks – do not rise and fall on a monthly basis. They increase in size at puberty and then retain their protruding shape until the shrinkage of the body in old age. They are at their firmest in young adult females, who transmit sexual rump signals regardless of whether they are ovulating or not. This reflects the fact that human copulation is no longer exclusively concerned with procreation, but is also literally a matter of making love.*

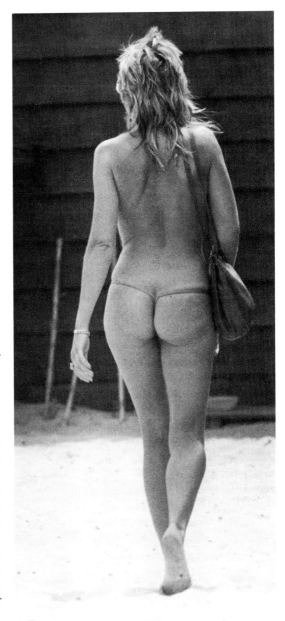

Today we may be seeing less of this female buttock signalling than was once the norm. It seems likely that the females of our early ancestors were, in fact, much bigger-buttocked than their modern counterparts. Evidence of this cannot, of course, be found in ancient skeletons, but when we look at Stone Age paintings and sculptures, huge buttocks are everywhere. They persist after the Stone Age in the prehistoric art of many cultures, but then gradually begin to disappear, dwindling to modern proportions which, although still relatively much larger than those of the males, are considerably less extreme. These early 'super-buttocks' have given rise to a great deal of speculation. One possible scenario runs as follows: Our primeval ancestors mated from behind, like other primates, so the pre-human sexual signals of the female came from the rear, as with other species. Then, as we evolved into the erect posture and our rump muscles bulged out into buttocks, the swollen shape became the new human sex signal. Females with larger swellings on their rumps sent out stronger sex signals, and this condition then started to increase until the buttocks became huge. The sexiest females had the advantage of supernormal buttock signals with their new super-buttocks, but these became so big that they actually began to interfere with the sexual act they were promoting. So the males solved the problem by switching to frontal copulation. As part of this new frontal approach, the breasts became permanently swollen as mimics of the large hemispherical buttocks. Now these super-breasts could also send out sexual signals, sharing the load, so to speak, with the buttocks, which could now decrease in size. This later version of the human female, better balanced and more agile, was at a considerable advantage over the fat-laden earlier model, which was gradually replaced.

■ *The difference in relative size between male and female buttocks acts as an important gender signal. The human rump has a protruding hemispherical shape in both sexes but is wider and more exaggerated in the female, a difference which becomes marked at adolescence. This is due to a widening of the pelvic girdle, connected with the demands of child-bearing, and to a heavier fat deposit on the female buttocks.*

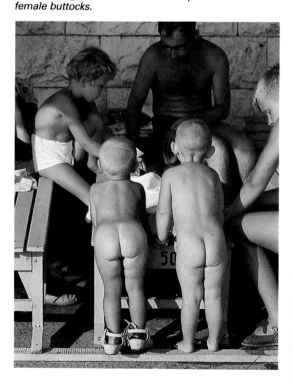

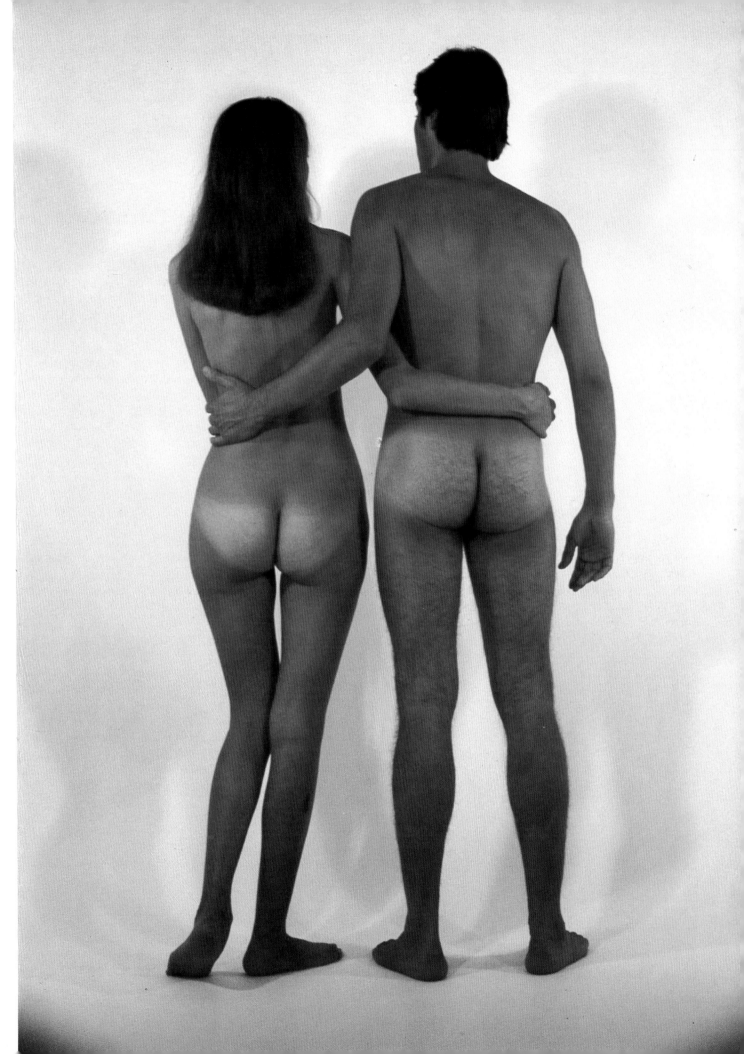

If this speculative sequence is correct, we would expect to find some remnants of evidence to support it. Those remnants are to be found today in the southwestern deserts of Africa, where Bushmen females still display the super-buttocks depicted in the Stone Age figures. Their remarkable contours reach astonishing proportions in some individuals and may well be showing us today what *all* our ancestral females used to look like, many thousands of years ago.

It has been argued that to compare Stone Age Europeans – who were presumably the models for the Stone Age figurines – with modern-day Bushmen living in the far south of Africa, is nonsensical, but this objection overlooks the true history of the Bushmen. They are not living in their remote desert because it is their favoured environment. They are there because it is the last corner of Earth where they have been able to cling on, as a disappearing branch of the human family. In earlier days their ancestors owned much of Africa and left their beautiful rock paintings behind to prove it. But they represented the Old Stone Age, the period epitomized by hunting and gathering as a way of life. With the arrival of New Stone Age peoples – the early farmers – they were driven from nearly all their territories, and today only some 50,000 survive, scarcely enough to populate a small city. Yesterday, however, they were one of the dominant forms of our species, and there is no reason to suppose that very large buttocks (the condition called steatopygia) was some kind of obscure desert rarity. It is more than likely that in the primeval hunting phase of human prehistory huge buttocks were normal for women and that the Stone Age artists based their figurines on reality rather than erotic fantasy.

When the slender, more agile females came to dominate the scene, the old big-buttock image did not fade away completely from the human unconscious mind. It still re-surfaces from time to time in rather unexpected ways. Many trivial costumes and dance movements exaggerate the buttock region. Even in staid Victorian times, with the introduction of the bustle, the male gaze was offered a new, artificial version of steatopygia. Hoops, padding, wire netting and steel springs were brought into play to re-create the long-lost fatty protrusions in the rump region. The elegant ladies who wore their bustles in polite Victorian society would no doubt have been horrified at such a view of their costumes, but today the comparison seems inescapable. In the twentieth century the major device for exaggerating the female buttocks has been the high-heeled shoe. This distorts female walking in such a way that the buttocks are thrust upwards and outwards more than normally and are forced to undulate even more when in motion.

Even without undue exaggeration the buttocks continue to provide one of the main erotic focuses of the modern female body. Long dresses that hide the legs are often cut so as to display the contours of the behind and clearly delineate their movements. Short garments like the 1960s mini-skirt display the buttocks more directly, and tight trousers, while obscuring the actual flesh, leave no doubt about the precise shape of the hemispheres.

In the early 1980s, there was a brief period of great emphasis on carefully designed, tight-fitting, high-priced jeans, deliberately intended as the perfect 'casing' for showing off this region of the body

■ *Because of our cultural taboo on the exposure of buttocks, some entertainers can earn a living simply by displaying them in an erotic theatrical context. For many years this was an exclusively female activity, but recently, with the increase in female sexual honesty following in the wake of the feminist movement of the 1970s, there has been an increasing interest in male strip-tease.*

■ *When a modern artist converts the female buttocks into 'super-buttocks', this is a further exaggeration of the trend that already exists in reality. In primeval times huge female buttocks were apparently commonplace, and the modern male may well carry a relic-response to them without realizing it. Almost all the surviving Stone Age figurines of human female figures show a condition of steatopygia (literally: fat-buttocks), and this fat-buttocked shape is still visible today in the Bushmen and Hottentots of Africa. It was strongly echoed in Victorian times – when the female buttock region aroused unusually intense interest – with the introduction of the bustle.*

as a bold sexual signal from the newly liberated female. The author of a booklet called *Rear View*, published at the time, and devoted exclusively to the erotic impact of female buttocks, hailed the new fashion era with these breathless words: 'The Butt Blitz began in 1979 when a spokesperson...thrust her vibrating, gyrating, designerized derriere into the startled face of network television... It was the start of the cultural phenomenon known as Designer Jeans.'

Inevitably within a few years designer jeans were out and baggy boiler-suit trousers were in, borrowing their shape to some extent from astronaut-wear, but the melodramatic language of the designer-jeans quotation nevertheless reveals the way in which the fashion world keeps on returning to the primeval buttock region as an erotic focus. We may have long since given up locomotion on all fours but the sexual rump refuses to fade from the unconscious mind of the male. It has even been suggested that the universal symbol of

love, the stylized heart shape, is in reality based on the buttocks. It certainly looks very little like the real heart, but with the cleft in its upper surface it does have an uncanny resemblance to female buttocks seen from behind. Here again a primeval human image may be at work.

In their newly liberated condition, females themselves have at last admitted that they too find buttocks appealing. Questionnaire studies have revealed that young females rate the shape and condition of male buttocks highly when judging sex appeal. There is a very strong preference for small, tight, muscular behinds, as might be expected since these are the essentially masculine features of this part of the human anatomy. One young actress has even gone to the length of publishing a whole book on the subject entitled *A Woman Looks at Men's Bums*. Full of close-up photographs of male behinds, it is a 1980s celebration of the newly gained sexual honesty of the modern female.

Up to this point we have looked at insulting buttocks and sexual buttocks, but there is a third way in which this part of the body has been displayed, and that is submissively. The presentation of the buttocks in a humble bent-over posture has had an enduring role as

The female buttocks, not the heart, is the probable source of the ♥ symbol.

■ *The traditional heart symbol, with its deep cleft, looks very little like a real heart. Unconsciously, it seems to have been based more on the shape of the naked female buttocks, as seen by an amorous male approaching from behind. As a symbol of love, this makes much more sense for the human species, bearing in mind the unique and highly characteristic shape of this part of the anatomy, so much the focus of male sexual attention.*

an appeasement gesture. In this respect there is no difference whatever between the behaviour of the submissive human individual and a submissive monkey or ape. In all cases, the 'presenter' is saying 'I offer myself in the passive female role. Please show your dominance by mounting me instead of attacking me'. Subordinate monkeys of either sex will make a rump presentation to dominant monkeys, also of either sex. The dominant individuals rarely attack such a subordinate, either ignoring it, or else mounting it briefly and making a few formalized pelvic thrusts. As an appeasement display the action is valuable because it enables a weak subordinate to remain close to a powerful dominant without being attacked.

In some tribal societies it has been observed that the bow performed as a greeting ceremony is done facing *away* from the greeted person. This looks so much like a 'rump presentation' that it is hard not to see it as related to the typical primate appeasement action. A

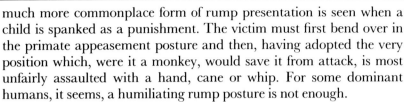

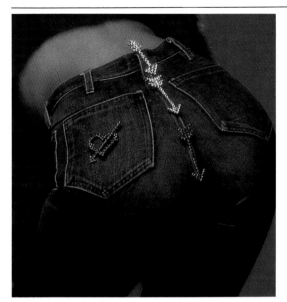

much more commonplace form of rump presentation is seen when a child is spanked as a punishment. The victim must first bend over in the primate appeasement posture and then, having adopted the very position which, were it a monkey, would save it from attack, is most unfairly assaulted with a hand, cane or whip. For some dominant humans, it seems, a humiliating rump posture is not enough.

Because of their sexual implications inter-personal buttock contacts are also somewhat restricted. Outside the sphere of loving couples, the pat or mild slap on the behind can only be used safely as a signal of friendship when there is no danger of sexual implications. Employed between friends at an ordinary social gathering it could

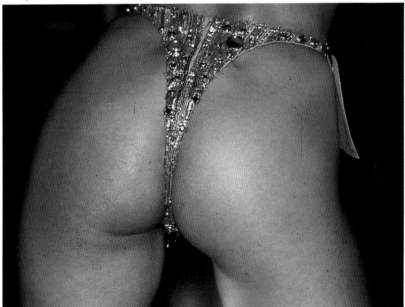

■ *Ornamented buttocks are rare. Jewellery cannot be attached because of the difficulty in sitting, but occasional tattooing can be seen. Even this is uncommon, however, because it is difficult for the owner to admire it; and its display to others is, of necessity, limited to a small audience of intimates. In exotic contexts there is one way in which something approaching a bejewelled condition can be achieved. This is the wearing of a richly decorated G-string that cuts its way through the cleft of the buttocks. For those clad in tight jeans there is the possibility of suggestive badges and emblems to emphasize the erotic features of the buttock zone.*

▲ Unlike the back embrace, the buttock embrace, with one hand clasping each cheek, is exclusively sexual and used only between intimate lovers. It is comparatively rare in public places, except in the most free and informal contexts.

▼ Spanking and whipping of the buttocks has a long history in sado-masochistic relationships. For the masochistic female the rhythmic beating of her rump region is like a painful version of savage pelvic thrusting by the male. For the sadistic male, the beating actions are like symbolic stabs of the penis. It has been pointed out that the reddening of the buttocks that results from this harsh treatment also acts as a sexual stimulus, mimicking the sex flush which occurs with intense arousal in the human species. In addition, the writhing and tensing of the beaten female resembles the actions of orgasm.

easily be misconstrued and the pat-on-the-back is preferred to the pat-on-the-behind unless sexual innuendos are deliberately intended. The pat-on-the-behind is therefore restricted to such contexts as a parent with a very small child, or sportsmen during violent team contests. In both these cases, sexual thoughts are so remote from the relationships that no misunderstanding can arise. By contrast, elderly relatives or 'friends-of-the-family' who exploit age differences by patting the bottoms of teenage daughters – enjoying mild sexual touching disguised as harmless pseudo-parental contact – can be the source of much annoyance.

Between lovers, buttock-clasping is common in both courtship and copulation itself. It is a frequent accompaniment of advanced stages of kissing and embracing, the back embrace being lowered to a buttock embrace as arousal increases. In old-fashioned ballroom dancing, where strangers are permitted to enjoy a frontal embrace as they dance, a male partner may exploit the situation by letting his hand shift down his partner's back towards her buttocks. In the classic film caricature of this strategy he quickly finds the offending hand returned to its original position.

During the advanced stages of copulation itself buttock-clasping often becomes quite powerful buttock-grasping as an accompaniment to vigorous pelvic thrusting. It is during this phase of body-contact that the hemispherical shape of the buttocks becomes so intimately linked in the minds of the lovers with intense sexual feelings.

It is this sexual linkage, again, that causes the occasional furore over the notorious Italian pursuit of public bottom-pinching. Any attractive girl walking the streets of an Italian city is liable to have her buttocks pinched by admiring strangers. According to her social background she may respond with pride, amusement, irritation or outrage. The author of a satirical work entitled *How to be an Italian*, describes bottom-pinching as an essential ingredient of the *Kama Pizza* and lists the following 'three fundamental pinches': 1 – *The Pizzicato*: a quick tweaking pinch performed with the thumb and middle finger. Recommended for beginners; 2 – *The Vivace*: a more vigorous, multi-fingered pinch performed several times in quick succession; and 3 – *The Sostenuto*: a prolonged, rather heavy-handed rotating pinch for use on 'living girdles'. Modern feminists have long since ceased to find this subject a source of humour and have even on occasion struck back by taking to the streets in search of male buttocks for concerted pinching assaults.

Finally, the buttocks as potential areas for body decoration provide little scope. They are too private for displaying the handiwork and too sat-upon for the attachment of ornaments. Tattooed buttocks are not common, except among the true fanatics. The only example of ornamented buttocks comes from the seventeenth century work *Man Transformed* by John Bulwer, in which he shows a particularly miserable looking native with jewels hanging from the left buttock. Bulwer comments: 'Among other filthy-fine devices of some Nations, I remember..a certaine people, who in an absurd kind of bravery, bore holes in their buttocks, wherein they hang pretious stones. Which by their leaves must needs prove but an inconvenient and uneasie fashion, and very prejudiciall to a sedentane Life.'

THE GENITALS

■ When our early ancestors first took to walking about on their hind legs they found themselves unavoidably offering a full-frontal display whenever they approached their companions. Previously it had been normal to advance on four legs, with the genitals completely concealed and well protected. To display the genitals then required a special posture. Now they were on show every time one human animal turned towards another. This meant that it was impossible for an adult to approach another without making a sexual statement. As a way of damping down these signals both males and females eventually took to wearing some kind of covering over the genital region: the loin-cloth was born.

The loin-cloth had three advantages. It not only reduced the strength of the sexual display of its owner when he or she was in a non-sexual public context, but it also intensified the sexuality of the private moments when it was removed. Thirdly, it helped to protect the delicate genital region from the harsher surfaces of the natural environment.

Today whenever people shed their clothes because of the heat it is always the modern equivalent of the loin-cloth that is the last to go. Unless we are dedicated nudists, we reserve our genital displays almost exclusively for our sex partners. Only for tiny children in a clearly pre-sexual phase do we relax this rule. In most countries we do not rely on custom alone to keep adult genitals covered but also impose formal controls: it is against the law to 'exhibit' one's genitals in public. Generations of pious church-goers have responded to calls from the pulpit – 'Nudism is...as shameless as the Devil himself... It is the zenith of human rebellion against God.'

What precisely are we at such pains to conceal? In the case of the adult male there is a triangular tuft of dense curly hair surrounding an unusually thick penis, behind and below which hang a pair of dangerously exposed testicles. In the adult female there is a similar tuft of pubic hair which surrounds and to some extent conceals a vertical genital cleft. In the standing female little more is visible, but seated with the legs apart she reveals a small nipple-sized clitoris and fleshy labia bordering her vaginal opening.

First, the pubic hair. In both sexes this appears at puberty, when the sex hormones come into operation. It acts as an immediate visual signal indicating sexual maturity and it has been suggested that this was its primary function during the early days of the human story when it first developed as a distinct patch of hair. It is certainly true that in pale-skinned races the dark triangle of hair is highly conspicuous even at a great distance. The argument is less convincing for dark-skinned people, however, and it seems likely that it had other functions as well. Two have been suggested. One is that the dense tufts of hair acted as buffers to prevent skin-chafing during the vigorous and often prolonged pelvic thrusting of face-to-face copulation. Bearing in mind that whole cultures have practised depilation of their pubic regions without any recorded skin damage, this too seems a little far-fetched. It is the third function that seems to be nearer the truth. The suggestion here is that, as with the tufts of hair in the armpits, the pubic tufts are essentially scent carriers. There is a powerful concentration of scent glands in the crotch region and dense hair there acts as a scent trap. As before, tight clothing can easily

■ In most cultures surviving today the genitals have become 'private parts' and a 'genital cover-up' is required in public places. The methods of concealment have varied considerably. One of the more remarkable was that employed in the 1970s by dancers in New Britain who, when performing for the British royal family, wore 'a modesty covering consisting of a tube of bark fitted with a disc and supported by a cord round the waist'.

create problems for this scent-signalling system, allowing the secretions to go stale and converting the natural sexual fragrances into unpleasant body odours.

At puberty, boys are always proud of their newly sprouting body hairs, but some girls are less enthusiastic. They are fully aware that their pubic hair is a badge of sexual maturation, but unconsciously they may feel that it is somehow masculine because the bodies of adult males are generally much hairier than those of women. As a reflection of this there is a quite irrational fear of 'hairy spiders' among females at the age of puberty. Among ten-year-olds there is no difference in spider reactions between boys and girls, but by fourteen spider hatred is twice as strong in girls as in boys, and the word 'hairy' is always added descriptively with a shudder.

It is not surprising therefore to find that females have frequently

■ We have no idea how soon the vertical posture of our species led to the invention of the loin-cloth as a means of reducing the sexual impact of the newly exposed genital region; but today, even where an atmosphere of sexual honesty and freedom prevails, this body zone is the last to be uncovered. In some tribal societies, like the Nuba in the Sudan, genital nudity is permitted, subject to certain restrictions. Only those who are young, completely healthy and physically handsome or beautiful, according to local standards, are allowed to go unclothed. During the annual love-dances the young females select a mate by lifting one leg and placing it on the shoulder of their intended one. This brings the girl's genital region into close proximity with the seated male's bowed head. Since he is not supposed to look at her, the implication is that she is transmitting an olfactory signal.

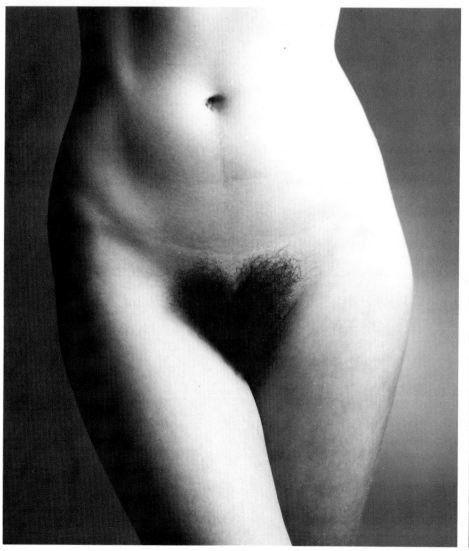

■ *The manner in which the female genital region may be displayed or depicted in public has always been hotly debated, and much importance has been attached to the presence or absence of pubic hair. The pubic hair of the human species is a visual sign of sexual maturity and also a sexual scent trap. It features prominently in the bawdy art of all periods but, because of its sexual significance and cultural symbolism, is omitted from public works of art in more prudish epochs and, as recently as the 1940s and 1950s, was removed from photographs of nudes by careful retouching. By the 1960s, however, restrictions had been widely relaxed and it was even possible to advocate the decorative trimming of the pubic triangle into a heart shape, as a new fashion fad.*

removed their pubic hair to enhance their beauty. In ancient civilizations this was done by painful plucking or, rather hazardously, by singeing or burning away the hairs, but in more recent times the favoured methods have been electrolysis, the application of depilating creams, or the shaving of the region using a small safety razor. In certain Moslem countries, however, the ancient plucking method is still used, little changed since the days of ancient·Rome and Greece.

There is a long tradition of pubic depilation among female artists' models and up until the present century all works of art (other than deliberately pornographic ones) showed the female figure totally devoid of any hint of pubic hair. So total was its suppression that innocent men reared by puritanical parents were quite unaware of its existence on real women. The famous art critic John Ruskin was so horrified to discover on his wedding night that his beautiful young bride had a hairy crotch that he was incapable of making love to her. He procrastinated for year after year until she finally lost her patience, had herself medically examined to prove her virginity, obtained an annulment of the marriage, and became the wife of the artist Millais.

It was not until the wild men of modern art, like Modigliani and Picasso, began to include pubic hair in their paintings that the centuries-long tradition was broken. Photography eventually followed suit, the earlier air-brushed portraits of nudes being replaced by frank, unretouched ones. Finally the cinema and in some countries even television permitted the occasional glimpse of a pubic tuft, although the potency of its sexual message still prevented it from becoming commonplace.

In the naked human female the pubic hair does tend to mask the genital details, and early writers who dared to discuss such matters were of the opinion that this was a wise strategy on the part of the Almighty. However, if God were attempting a cover-up he failed dramatically with the human male. Both the penis and the testicles, although heavily fringed with hair, remain stubbornly conspicuous. They are also surprisingly vulnerable to injury and the testicles in particular seem to be wantonly placed in a position of maximum risk. Why any human organ so delicate and unprotected should be brought down to hang loosely between a pair of powerful, active legs is at first sight mystifying.

The human testicles start out life tucked away in the abdomen, but two months before birth they descend through an opening called the inguinal canal and take up their external position. There they dangle for the rest of life in the cooler world outside. It is this drop in temperature which appears to be the crucial advantage in their move. Inside the body they would have been kept at 98·4°F, the normal body temperature, but in the outside air they manage to lose a few degrees. This is essential because the sperm-production cells inside the testicles cannot operate efficiently at the higher temperature. There is even some fine-tuning to this air-conditioning system: when a male takes a very hot bath or a sauna the suspension cord lengthens so that the testicles are farther away from the body; if he takes a cold shower the cord shortens rapidly, elevating the testicles until they are in close contact with the warmth of his crotch. If a man suffers a fever and runs a very high temperature for a few days, he may even become

▲ This Pygmy fisherman going about his daily business clearly demonstrates that there is no predictable link between large, powerful male build and large penis size. If anything there seems to be an inverse connexion, with small slender males often having proportionately larger genitals than their otherwise more-amply-endowed friends.

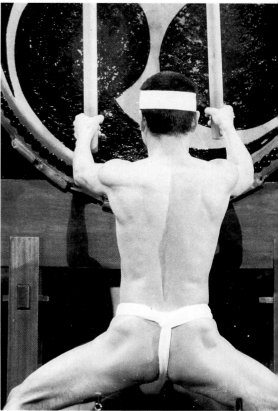

temporarily sterile. In some tribal cultures the men make use of this temperature sensitivity as a primitive means of birth control, dipping their testicles in very hot water for several days before copulating. A word of warning is necessary, however. A group of volunteer medical students who exposed themselves to high sauna temperatures eight times in two weeks and then had their sperm count checked for some weeks afterwards did not show the expected dramatic drop in sperm production. There was a slight reduction but it was far less than expected. So the system is extremely unreliable. Brief periods of exposure to heat are insufficient. Clearly only very prolonged high temperatures destroy sperm production, as in the case of a severe fever lasting for several days.

Cold is not the only factor which elevates the testicles. They are also pulled up tight against the body during fear, anger and sexual arousal. The freezing man's response is purely heat-seeking, but fleeing, fighting and fornicating men all share one and the same need to protect the testicles from physical damage. In fleeing and fighting the increased risk is obvious enough, but the 'sexual elevation' is slightly more surprising, especially as the bodies involved are flushed with heat. The explanation seems to be that during the most intense

■ *This 18th century painting of love-making by Thomas Rowlandson and the 1980s Californian street scene showing a woman eating a phallic ice-cream both reflect a liberated attitude towards human sexuality. Puritans have frequently sought to reduce human sexual activity to the level of purely procreative behaviour, insisting that any other kind of sexual indulgence is unnatural. The biology of the human species refutes this, the fact that females no longer signal their moment of ovulation to their males being especially significant.*

◄ *Tight clothing enables males to transmit a gender signal from the genital zone even though it is covered. The 'crotch-bulge' caused by the external male genitals contrasts strongly with the 'crotch-gap' of tight female clothing. The impact of this signal has frequently been exploited by clothing manufacturers in their advertisements and in the design of minimal male costumes in theatrical productions.*

moments of copulation, when the mating pair are at their most abandoned, there is such vigorous pelvic thrusting taking place, with the imminent possibility of threshing and writhing limbs during female orgasm, that an orderly testicular retreat was the appropriate step for evolution to take.

A strange feature of the sexual elevation is that the testicles become temporarily enlarged. During moments of peak arousal these vaso-congested organs will have increased in size by 50 per cent and some-

During orgasm the brain is flooded with a chemical that stimulates the pleasure centres.

times as much as 100 per cent. Another odd feature of the human testicles is their asymmetry: in 85 per cent of males, the left testicle hangs slightly lower than the right. Why this should be is not clear. It means that during elevation the left testicle has to rise farther, because eventually they both come to the same raised-up level.

The worst aspect of the external positioning of the testicles is the opportunity it provides for the easy castration of males. This has been done as an act of crude brutality, as a formal punishment, as a method of creating a eunuch class of males who could not reproduce, and as a way of creating adult male sopranos.

The Vatican habit of removing the testicles of small boys in order to keep them singing on as high-pitched adults must have dramatically increased the number of tone-deaf children in Rome. The castrati were a feature of the Christian church for many centuries and it comes as a surprise to discover that the operation was routinely performed until as recently as 1878, when an enlightened Pope finally outlawed it. Testicles seem to have been something of a preoccupation of the Vatican, because at one time there was a special chair built, called the Porphyry Chair, with a horseshoe-shaped seat on which a new Pope had to sit to have his private parts inspected by cardinals to ensure that the Holy Father was not a Holy Mother or, for that matter, a Holy Castrato. The cardinals, having confirmed the presence of the Papal scrotum, had to proclaim in Latin: 'He has testicles and they hang well.'

If you are an adult male reader and you say these words: one hundred, two hundred, three hundred, four hundred, five hundred – the time it will have taken you to do so will have been sufficient for your testicles to have created another 15,000 sperm. This astonishing rate of production goes on all day and every day. We have no breeding season like other animals – at the level of spermatogenesis we are constantly sexually active, producing sperm at the rate of 3000 a second. If young males who are at the peak of their sexual development do not find sexual outlets during their waking hours they may experience spontaneous ejaculations during their sleep. If this does not happen, then the unused sperm are absorbed back into the body.

If sexual stimulation leads to a male orgasm, between 200 and 400 million sperm are ejaculated through the erect penis. Even this huge number would cover no more than a pinhead, but they are not sent out into the world unprotected. They are immersed in a blob of seminal fluid, produced by the prostate gland and the seminal vesicles. The fluid contains special proteins, enzymes, fats and sugars to nourish the sperm, and an alkaline liquid medium in which the sperm can swim. This alkalinity is vital because the female's vagina is acidic and needs to be neutralized if the sperm are to survive their crucial journey.

The average amount of seminal fluid produced for each ejaculate is 3·5 millilitres – about a teaspoonful – but healthy young males who have been sex-starved for a long time can produce up to four times as much. There are several ejaculatory spasms, usually three or four, at fixed intervals of 0·8 of a second, which can launch the seminal fluid a surprising distance. Although it is not one of the figures mentioned in the Guinness Book of Records, intrepid research workers have measured a three-foot ejaculation; 7 to 8 inches is more usual.

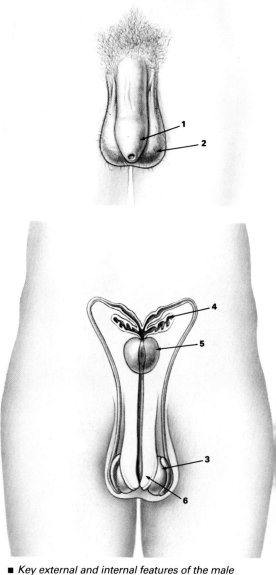

■ *Key external and internal features of the male reproductive system are: the penis – complete with foreskin (1), in this instance; the scrotum (2); the testicles (3); the seminal vesicles (4); the prostate gland (5); and the glans of the penis (6).*

The human penis is remarkable for its large size and its lack of a supporting bone.

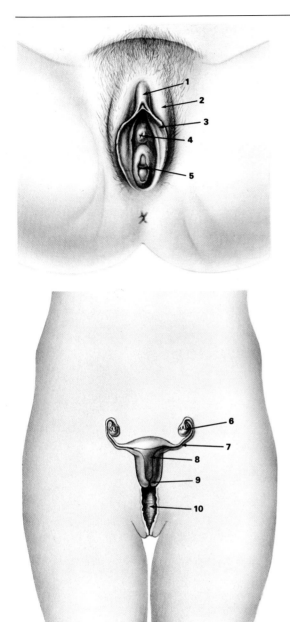

■ *Among the main external and internal features of the female genitals are: the clitoris (1); the outer labia (2); the inner labia (3); the urethral opening (4); the vaginal opening (5); the ovaries (6); the fallopian tubes (7); the uterus, or womb (8); the cervix, or neck of the womb (9); and the vagina (10).*

It is crucial for the survival of the species that the sensation accompanying ejaculation should be an intensely pleasurable one so that, whatever the obstacles, men return again and again to this climactic action. This is not left to the higher centres of the brain but is dealt with at a very basic chemical level. During orgasm the brain gets flooded with a natural chemical product, very similar to morphine, which kills pain and stimulates the pleasure centres. At the same time part of the brain is firing off its nerve-cells in a manner remarkably like someone having a brief epileptic fit, while the higher centres are almost completely switched off for a few magical moments of abandon. Just before all this happens the vision centres receive a sudden blast of intense stimulation giving the impression of a flood of light. To sum it all up, the male orgasm is rather like a momentary, natural 'drug high', with pain and worry banished and a vertiginous pleasure suffusing the system, blotting out the external world. It is little wonder that the planet has become so overpopulated.

The human penis, the sperm-delivery organ, is remarkable for two things: its large size and its lack of an *os penis* or supporting bone. Its size is greater than anything seen in other primates. Even the mighty gorilla, weighing three times as much as a man, has a much smaller appendage. The relaxed human penis has an average length of 4 inches, diameter of 1¼ inches and circumference of 3½ inches. When fully erect it has an average length of 6 inches, diameter of 1½ inches and circumference of 4½ inches. Variation in penis size is quite considerable, the largest erect one recorded having been measured at 13¾ inches – more than double the size of the average one. Strangely there is no connexion between overall body size and penis size, the largest organs often occurring perversely on the smallest, skinniest males. The angle of the erect penis also varies, but in the average male it is about 26 degrees above the horizontal, which matches up well with the angle of the female's vagina.

The penis consists of a glans, shaft and root. The name glans means 'acorn', the slightly enlarged head of the penis having a roughly hemispherical shape. When the penis is relaxed the glans is completely covered by a loose sheath of skin called the prepuce or foreskin. This protects the delicate tip of the penis and helps it to maintain its sensitivity for those special moments when erection exposes it to contact with the female genitals.

The rounded softness of the glans, even when the penis is erect, helps to avoid pain or damage to the female surfaces during copulation. This is important because of the unusual width of the human penis. By comparison, a monkey penis is a thin spike supported and stiffened by a small bone, the *os penis* or *baculum*. The thickness of the human organ makes this bone unnecessary, sufficient stiffness being obtained by the engorgement with blood of the spongy tissues of the penis shaft. This engorgement, which can create a full erection in as little as five seconds, has been described as a 'triumph of mechanics which no engineer would have thought possible'. It operates on a delicate balance between increased input of blood and decreased output. There is a blood 'traffic jam' of just the right intensity to turn the penis into a hard rod suitable for penetration of the vagina without causing the male any pain and without any sudden fluctuations during sexual arousal. And it has to accomplish this without blocking

Male circumcision as a tribal badge has very ancient origins.

the urethral tube which runs through the spongy blood-filled tissues – the tube which must deliver the seminal fluid through the *meatus*, the urethra's external orifice.

The reason for this unusual penis design – thick and boneless, with complex vaso-congestion – seems to be connected with the giving of sexual pleasure to the human female. Female monkeys receive a few rapid thrusts from the thin bony spikes of their males and in a trice copulation is over. Female monkeys do not enjoy the explosive orgasms experienced by the human female. The thick human penis causes powerful contact sensations as it moves against the surfaces of the female genitals, during the often-lengthy pelvic thrusting of our species. The female orifice, surrounded by highly sensitive folds of skin, the labia, is subjected to repeated, rhythmic massage by the tightly fitting penis. As female sexual arousal mounts, both the outer labia and the inner labia become engorged with blood, swell to twice their normal size and develop greatly increased sensitivity to touch. After prolonged stimulation the female eventually experiences an orgasmic climax which is physiologically very similar to that of the male. This means that both partners experience a massive reward for their sexual labours and the encounter, unlike that of monkeys, may lead to a tight emotional bonding between the pair. The fact that the female gives no clear signal to the male when she is ovulating also means that the majority of copulations are not procreative but instead serve to further tighten the emotional bond.

Just above the orifice of the vaginal tube is the small nipple-sized female equivalent of the male penis – the clitoris. This organ is purely sexual in function and also becomes enlarged and more sensitive during copulation. During foreplay it is often stimulated directly by touch and during pelvic thrusting it is massaged indirectly but vigorously by the movements of the inserted penis, which drags rhythmically on the skin in its vicinity.

The vaginal passage is about 3 inches in length. This dimension becomes only slightly increased during intense sexual arousal, so even a modest penis is more than able to reach its far end, where the sperm are ejaculated against the cervical opening. Through this they swim on their great journey across the uterus to the fallopian tubes where, if the timing is right, they will meet a minute egg descending, and one of them will join with it to start a new life.

Although the female's ovaries contain literally thousands of eggs, she sheds no more than 400 during her reproductive lifetime. They mature at the rate of one a month and are fertile only as they pass down the fallopian tubes, a four inch journey which takes them several days.

These then are the human genitals. Considering their great delicacy, complexity and sensitivity, one might imagine that an intelligent species like man would leave them alone. Sadly this has never been the case. For thousands of years, in many different cultures, the genitals have fallen victim to an amazing variety of mutilations and restrictions. For organs that are capable of giving us an immense amount of pleasure they have been given an inordinate amount of pain.

The commonest form of assault they have suffered is male and female circumcision. This strange mutilation is older than civiliz-

■ *Male circumcision is practised in many tribal cultures around the globe and appears to have a very ancient origin. It was common in Egypt at least 6,000 years ago, from where it seems to have spread to both Muslims and Jews. In recent times it has acquired medical respectability and been widely performed on Christians and others for whom it has no religious significance whatever. In fact circumcision has no medical advantages, and some disadvantages. Suggestions that retaining the foreskin may cause certain kinds of cancer have now been proved false. The significance of circumcision in the past has not been medical but social. It has acted as a form of 'loyalty oath' – an ordeal shared by young males who are then more closely bonded to one another by their common suffering. In western society today this significance has largely been lost because the mutilation of the penis is almost entirely restricted to tiny babies who are quite unaware of the symbolic oath of loyalty which they are being forced to take.*

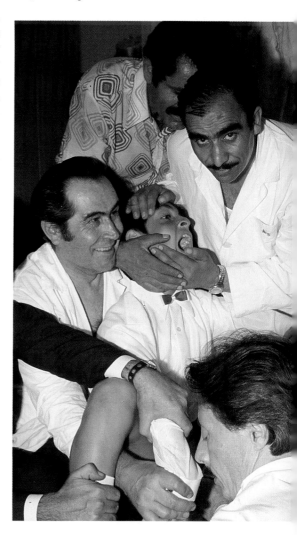

▲ *One of the arguments calculated to make circumcision medically respectable involves the suggestion that many male babies are born with a foreskin that is too tight and will therefore not retract properly during sexual activities in the future, when they are adult. Snip now to avoid embarrassment later is the motto. But this is a misinterpretation of the biological facts. The truth is that the vast majority of babies have foreskins which will not retract fully. This is the normal, natural condition of the infantile penis. In most male children it takes several years for the retractability of the foreskin to develop.*

■ *Female circumcision is much more serious a mutilation than its male equivalent. Removal of the male foreskin has little effect one way or the other on sexual performance and pleasure, but removal of the clitoris and the labia of the female does drastically reduce sexual responsiveness. This, in fact, is the true significance of the female operation, rendering the circumcised female more easily controlled as the 'property' of her male. Many millions of females alive today in parts of Africa and the Middle East have had to endure the agony of having all or some of their external genitals cut away.*

ation and was probably already well established in the Stone Age. Although it is a piece of deliberate wounding of children by adults, it has always been done with the best of intentions. Over the millennia it has caused countless deaths from infection, but its advantages have always been said to outweigh the risks involved. These alleged advantages have varied from epoch to epoch and culture to culture, but recent re-examination of the evidence has shown that they are all imaginary.

It has been claimed that one of the oldest reasons for performing male circumcision – the removal of the foreskin – is that it provided immortality in the shape of life after death. This odd notion was based on the observation that when the snake sheds its skin it emerges with glistening new scales and is 'reborn'. If the snake can enjoy rebirth by the removal of skin, so too can the human being. For snake read penis; for snakeskin read foreskin.

Once male circumcision had become traditional it no longer mattered whether the old beliefs survived. Being circumcised was now a badge of belonging to a particular society. The ritual mutilation spread and spread. Ancient Egyptians were doing it as long ago as 4000BC. In the Old Testament, Abraham demanded it. Arabs circumcised as well as Jews. Mohammed was said to have been born without a foreskin (which he may well have been, as this condition is not unknown to medical science), a claim which automatically doomed the foreskins of his future male followers.

As the centuries passed, religious reasons gave way, for many, to quack medical arguments. The possession of a foreskin was said to cause 'masturbatory insanity'. Other medical horrors resulting from its retention included: hysteria, epilepsy, nocturnal incontinence, and nervousness. Such ideas survived into the early part of the present century and even led to the formation of an Orificial Surgery Society devoted exclusively to the 'modification' of offending genitals as a means of preventing mental illness.

When at last this nonsense was on the decline a crisis arose. What new reason could be found for mutilating children's genitals? The solution had to be one that suited the rational climate of twentieth-century scientific enquiry. The answer appeared in that learned journal *The Lancet* in 1932: foreskins caused cancer! By the end of the 1930s 75 per cent of boys in the United States were being circumcised; by 1973 it was 84 per cent; by 1976 it was 87 per cent. Cancer had become the secular version of hellfire and brimstone, the perfect weapon for the anxiety-makers of a post-religious society.

To be more precise, the claim was that the 'debris' called smegma which collects under retained foreskins could cause cancer of the penis and also cancer of the cervix of the wives of the uncircumcised. The paper which started this false rumour was founded on faulty statistics, but nobody minded because here was a plausible new reason for slicing away at the infantile penis. Subsequent experiments, however, revealed that there is nothing remotely carcinogenic about the smegma produced under the fold of the foreskin, but they were widely ignored. Other investigations showed that women whose uncircumcised husbands always wore condoms were no more or less likely to develop cervical cancer than those whose husbands never wore condoms. But, again, nobody wanted to know. In one

▲ *Two scenes depicting the circumcision of adult males – one unwilling, the other compliant – from the ancient Egyptian city of Saqqara.*

The still-widespread custom of female circumcision is a cause of untold misery.

project, a country where there was no circumcision at all was compared with one in which all males were circumcised. The results showed, to the relief of the foreskin snippers, that prostate cancer was higher in the uncircumcised country. Unfortunately this form of cancer is an ailment of elderly men, and when a correction for age distribution was made the figures showed that this disorder was actually more likely in the circumcised country.

Not only was the cancer scare completely without foundation, but the operation of foreskin removal continued to prove a distinct health hazard for small babies. There were many cases of haemorrhage, ulceration of the urethra, surgical trauma and local infection. In rare cases foreskin removal resulted in the death of the baby. There were also more subtle effects with possible long-term implications: following circumcision male babies showed an increase in the level of hormones related to stress; sleep patterns altered; there was more crying and more irritability.

Despite all this, 'medical' circumcision continued (and still continues) at a merry pace in certain countries where private medicine is the rule. In Britain, significantly, there was a dramatic decline in the operation following the introduction of the National Health scheme and free treatment. It is impossible to refrain from asking why it should be that the operation sank to a level of less than one per cent (only 0.41 per cent of male babies in 1972) in a country where there was suddenly no financial gain to be made from it, while in the United States, for example, in the same year, over 80 per cent of male babies were circumcised, at an annual cost to health-insurance companies of more than $200 million. The new deities demanding foreskins appear to be more fiscal than sacred.

Young females have also been assaulted in a similar fashion. This has been rare in the West, although as recently as 1937 a Texas doctor was advocating the removal of the clitoris to *cure* frigidity. The harshest traditions of female circumcision are found in Africa, parts of the Middle East, Indonesia and Malaysia. It is a staggering fact that, far from being an ancient memory, the practice of cutting away all or part of the external genitals of young females is still going on in more than 20 countries.

No fewer than 74 million women alive today have been subjected to this mutilation. In the worst cases they have had their labia and clitoris scraped or cut away and their vaginal opening stitched up with silk, catgut or thorns, leaving only a tiny opening for urine and menstrual blood. After the operation the girl's legs are bound together to ensure that scar-tissue forms and the condition becomes permanent. Later, when they marry, these females suffer the pain of having their artificially reduced orifices broken open by their husbands.

The effect of this practice is to dramatically reduce sexual pleasure for wives in the countries concerned, which may be its hidden significance. A side-effect is a high number of deaths and serious illnesses caused by the unhygienic conditions under which the operations are performed, especially in such countries as Oman, South Yemen, Somalia, Djibouti, Sudan, southern Egypt, Ethiopia, northern Kenya and Mali. The continuance of such practices in the twentieth century against a background of modern enlightenment is clearly going to puzzle historians of the distant future.

THE LEGS

■ The long straight human leg accounts for half the body's height. When artists are sketching the human form they divide it up into four roughly equal parts: from the sole to the bottom of the knee-cap, from the knee-cap to the pubic region, from the pubic region to the nipples and from the nipples to the top of the head. This is the adult shape. The proportions of children are slightly different, with the legs being shorter in relation to the upper body.

The foundation of our powerful legs comprises four bones: the massive thigh bone, the longest bone in the human body, called the femur; the knee-cap which protects the front of the hinge joint at the base of the femur, called the patella; the shin bone, which articulates with the femur, called the tibia; and the splint bone, which lies along-side the tibia, called the fibula.

Propelled by its well-muscled legs, the human frame has sailed seven feet eight-and-three-quarter inches (2·35 metres) up into the air and has managed a long jump of twenty-nine feet two-and-a-half inches (8·90 metres). Marathon dancing has dragged on, week after week, with the participants in a state of near exhaustion, for as long as 214 days. Such feats of strength and endurance are a remarkable testimony to the evolution of the human legs during a million years of chasing and hunting. No wonder the legs have come to be thought of as symbols of stability, power and nobility.

Legs are also conspicuous in erotic contexts. Because adults possess legs which are both relatively and in absolute terms longer than those of children it is inevitable that long legs should be equated with sexuality. A curious description of a sexually appealing female which is often heard is that 'her legs reached up to her armpits'. In Western culture the exposure of the legs has long been used as a titillation device by adult females. At different periods the amount of leg flesh visible to male eyes has varied considerably. In the last century legs disappeared altogether for long periods and even a brief glimpse of an ankle was considered shocking. So intense and complete was this suppression of the 'erotic leg' that even the word itself became prohibited in polite circles. In the United States legs were called 'limbs'. Even the legs of tables and pianos were referred to as 'limbs'. Piano legs, since they appeared naked on platforms at concerts and supposedly conjured up images of naked human legs, had to be clothed in frilly 'trousers'. Other euphemisms for legs included 'extremities', 'benders', 'underpinners' and 'understandings'. At table a chicken leg became 'dark meat'.

It is hard for us to comprehend a social climate in which such extremes of prudery could flourish, but the fact remains that legs were a taboo subject for a very long time. Only after the First World War did they emerge from hiding, and even then they caused many a raised eyebrow. The rebellious young females of the 1920s were boldly exposing their calves and even their knees and this was too much for some men. They insisted that the new fashion was causing a decline in moral standards and that the 'modern girl' was behaving like a harlot. There were many cases of employees being forbidden to wear the new, shorter skirts to work. One man, described as a distinguished lawyer, complained that 'The provocation of silken leg and half-naked thigh...was devastating and overwhelming'.

The main significance of such comments is that they reveal the ex-

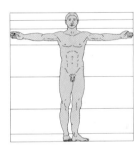

■ Human legs, in relation to the rest of the body, are the longest of any of the primates. The lower leg is roughly one-quarter of the total body height. So is the upper leg, which contains the largest bone in the human body – the femur.

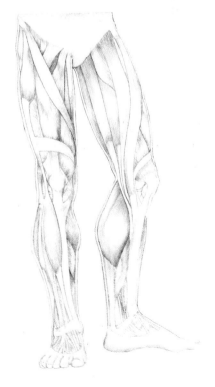

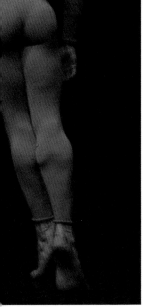

The legs as symbols of sexuality and strength.

■ The typical urban-dweller of modern times has spindly legs compared with his primeval hunting ancestor, but regular athletic exercise can quickly convert the scrawny limbs of desk-bound male office workers into muscular machines capable of greater speed, strength and staying power. Even the weaker legs of females can be built up impressively.

■ Because legs attain their maximum length at the time when sexual maturity arrives, it follows that extra leg-length symbolizes super-sexuality. Certainly, cartoonists nearly always employ a little artificial leg-stretching when creating pin-up girls. And it is noticeable that Miss World contestants always have slightly longer legs than the average female. The impact of their legginess is aided by the wearing of high-heeled shoes, which gives them a few extra inches in their struggle for an enhanced sexual shape.

tremely potent sexual signalling which emanates from young female legs. The reason is obvious enough. The more of a pair of legs which becomes visible, the easier it is to imagine the point where they meet. It would be an error to conclude from this, however, that changes in skirt-length during the twentieth century reflect nothing more than fluctuations in the sexual vigour of society. If we look at the rise and fall of the skirt, decade by decade, it is clear that short skirts arrive in periods of economic buoyancy and long skirts reappear during periods of economic decline. The short skirts of the roaring twenties were replaced by the long skirts of the depressed thirties; the long skirts of the austere post-war period in the late forties were replaced by the tiny miniskirts of the swinging sixties. These in turn made way for the long skirts of the seventies' recession. It is as if young females, influenced by the mood of society, reveal their level of optimism and self-confidence by the level of their hems. To the extent that an optimistic attitude goes with a lively sexuality it can be said that shorter skirts do reflect a society with greater sexual energy, but this is clearly only part of the story. The longer-skirted phase in the seventies, for example, was certainly not an outcome of prudishness.

The fact is that both the short skirt *and* the long skirt have sexual potential, with regard to the exposure of the legs. The short skirt has the advantage that it shows off the 'lower limbs' all the time, so that they are repeatedly displayed to males; but it also has the disadvantage that familiarity exhausts the male response. As any strip-tease dancer knows, you always start your act fully clothed and it is the slow removal of the skirt that makes the emergence of the legs such a strong sexual stimulus. The long skirt therefore has the advantage that it can make a powerful impact when raised or removed, but the disadvantage that for much of the time it blocks sexual signals from the legs.

What the very short skirts have symbolized, more than any sexual factor, is a sense of freedom. Females in short skirts can stride and leap and step out in the world. Those in long flowing skirts or tight tubular ones are engulfed in them and held back by them. The explosion of leggy mini- and micro-skirted girls in the sixties was the result of the new-found freedom stemming from the invention of the contraceptive pill and from the boom economy. The long legs transmitted the social message: 'We young females are on the move.'

By the time the eighties had arrived it was clear where the move had taken them – to the feminist movement and a renewed struggle for true sexual equality. With this last step came another shift. While the confused economic picture gave rise to confused fashions for skirts – some long, some medium and some short – the avant garde of the female population were side-stepping the issue completely by switching to leg-equality: they donned male leg-attire – jeans, slacks and trousers. These garments, which like short skirts had caused uproar when first introduced and which led to young females being thrown out of elite gatherings, quickly became acceptable in more and more contexts.

Just like short skirts and long skirts, tight female trousers had an advantage and a disadvantage. They revealed to the naked eye for the first time the precise shape of the region where the left leg meets the right. This gave them a strong erotic potential. But at the same time they interfered with the smooth shape of the leg, adding un-

■ *Because of the sexual significance of long female legs, their deliberate exposure has often been used as an erotic display. Employed today, this is a curiously antique signal, based on a pretence that legs are normally still covered as they were during periods when full-length skirts were insisted upon by polite society. Having lost much of its impact, the lifting of the skirt has now descended to the level of joke sex and parody.*

aesthetic folds and crinkles to the gently curving contours. They also gave the impression of a protective coat of armour, encasing the legs and robbing them of their vulnerability to male approach. In the mind's eye to lift a skirt is easy, to remove a pair of jeans is a struggle.

By and large, male legs have attracted far less attention than female legs, but that may simply be because comparatively loose trousers have been the dominant male fashion for so long. When, in earlier times, tight male hose was in vogue there was even a phase during which skinny-legged males could be seen strapping on false calves to give the impression of bulging muscles. It has nearly always been the case that the ideal male leg is the strong muscular one, and the advent of tight male jeans in recent years has once again meant that spindly-legged or flabby-legged males have been in difficulties. Luckily for them, loose trousers of one kind or another have always been available as alternatives.

Apart from their general shape and exposure, the human legs have also transmitted a variety of signals from their postures. There are three basic positions: legs apart, legs together and legs crossed.

Standing, sitting or lying with the legs wide apart are postures that signal stability, confidence and sexuality. Standing with the legs apart and the feet planted firmly on the ground is a relaxed, balanced position often adopted by dominant individuals. It ostentatiously makes no attempt to protect the vital genital region, implying that because of high status it expects no threat. Sitting with the legs apart is rather different. There is still a dominant, confident mood to the posture, but because the genital region is now pointed forward instead of downward, the sexual display element is much stronger. If the spread-leg sitters were naked this would be obtrusively obvious, but even when fully clothed the almost defiant sexual nature of the posture still makes its impact. Lying down with the legs wide apart makes a similar but slightly less powerful statement.

Inevitably, etiquette books have instructed their readers to avoid the parted-legs posture. Amy Vanderbilt, as recently as 1972, found it necessary to inform American women that it is 'graceful to sit with the toe of one foot drawn up to the instep of the other and with the knees close together.' All legs-together postures, whether standing or sitting, have about them an air of formality, politeness, primness or subordination. The soldier standing to attention, the sportsmen lined up for their National Anthem, the children posing for their school photograph, and the 'proper' young lady sitting decorously on her chair at a social gathering – all these are legs-together people. It is the essential neutrality of their leg postures that gives them an air of inhibited 'correctness'. By avoiding any specific or positive leg posture they are, in effect, saying: 'We want to avoid giving offence.'

The third basic position, that of crossed legs, has about it an air of informality. In the nineteenth century, women in polite society were forbidden to adopt this posture in public and even today the stuffier books of etiquette still disapprove of it. Here is Amy Vanderbilt, doyenne of modern American manners, again: 'Crossing the legs is no longer considered masculine in women, but there are good reasons to avoid it as much as possible. First...it creates unattractive bulges on the leg and thigh crossed over. Secondly, when skirts are worn short, crossed legs can be indecent or at least immodest. Thirdly it is said to

■ *The erotic impact of pointedly exposed female legs depends on the context. If, for example, the female in question is required to exhibit her legs while standing or walking on a raised platform, she is automatically trans formed into a slave-girl being offered for sale at a slave market. This degrades her in a way that certain male egos find irresistible.*

The particular eroticism of the female thigh.

encourage varicose veins by interfering with circulation.' She goes on to caution against the dangers of crossing one's legs when applying for a job, arguing that the informality of the posture may create an impression of immodesty, or that one is too casual.

The basis of this mood difference between the prim and proper legs-together and the relaxed and casual legs-crossed is the degree to which they indicate a readiness or an unreadiness to rise from the comfortable sitting posture. The legs-together position shows a deferential readiness for action. The legs-crossed position indicates that the sitter has 'settled in' and has no intention of suddenly leaping up in an attentive fashion.

Looking more closely at the action of crossing one leg over the other, it emerges that there are nine ways in which this is done. They are as follows: 1 – The Ankle-ankle Cross. This is the most modest and formal of the crossing postures. The amount of cross-over is very small and the position is only slightly removed from the formal legs-together. 2 – The Calf-calf Cross. This is not a common variant. It has a similar mood-flavour to the ankle-ankle cross – formal and 'correct'. These first two versions of the leg-cross are the only ones displayed by certain high-status individuals on public occasions. The Queen of England, for instance, has never been photographed with her legs crossed above the calf. 3 – The Knee-knee Cross. This is the first of the truly informal postures and is the most common one seen on ordinary social occasions. For females wearing skirts this is one of the actions which may lead to unintended thigh exposure. It is therefore available for (conscious or unconscious) sexual displays. 4 – The Thigh-thigh Cross. This is a more extreme version of the last one, in which the legs are crossed as tightly over one another as possible. Because of the design of their (wider) pelvic girdle, this particular posture is much more commonly performed by women. 5 – The Calf-knee Cross; 6 – the Ankle-knee Cross; 7 – the Ankle-thigh Cross. These three related postures involve the pulling up of one leg high on the other. It is a form of leg-crossing which, if performed by a female wearing a skirt, would expose not only her thighs but even the region of her crotch. It is therefore almost entirely limited to males and to the occasional woman wearing trousers. Because of its masculine bias it is particularly favoured by macho males who wish to emphasize their gender. It is noticeable that in the United States this type of leg-crossing is much more common than in Europe. The European male's typical knee-knee cross seems slightly effeminate to some American males, just as there is something faintly swaggering about American males adopting this posture when observed by European males. Furthermore, within America, the frequency of this type of leg-crossing increases dramatically as one approaches what might be loosely called 'cowboy territory' in the west. 8 – The Leg Twine. In this form of crossing, one leg is twisted around the other and held there by the entwined foot. This is a largely female posture. 9 – The Touching-foot Cross. In this special type of leg crossing, the crossed-over foot comes to rest alongside the calf of the other leg. This is an exclusively female posture, the action involved being uncomfortable for a male because of his pelvic design.

These forms of leg-crossing appear repeatedly at almost every informal social gathering and represent a form of body language that

■ *Exposure or decoration of the thigh region of female legs is highly erotic because it focuses attention close to the point where the two legs meet – the primary sexual zone. The strength of this signal can be increased by deliberately covering up the lower regions of the legs.*

Leg postures while seated involve an element of genital defence and genital display.

transmits subliminal mood signals from person to person. Apart from the gender signals already mentioned, they can be used to signal like-mindedness between friends. If two people think alike on a particular issue, then they are highly likely top adopt similar forms of leg-crossing as they sit together talking. If one person is much more dominant than the other, however, and is asserting his status, he will almost certainly adopt a different type of leg-crossing from his subordinate. The legs transmit the unspoken message: 'I am different from you.'

When people are sitting side-by-side, the direction of their leg-cross is also significant. If they are friendly they point the top leg towards their companion. If they are unfriendly this leg points away and helps to bias the body in that negative direction.

There is one final, significant element in leg-crossing, and this has to do with how tightly the crossed-over legs are clamped together. In general, it is safe to say that the tighter the cross, the more defensive is the mood of the person concerned. The legs-apart posture discussed earlier revealed a basic confidence in the performer. In a sense legs-crossed is the opposite of legs-apart and it has been suggested that, because of this, all people with crossed legs are defensive. This is an oversimplification, because many people feel more comfortable in a crossed posture and adopt it even when they are alone. But it is true to say that when someone feels ill at ease in company they are likely to clamp their legs together rather more forcefully than when they are completely relaxed, and this element of their posture does not go unnoticed, even if their companions are unconscious of their reactions. The leg-twinings and thigh-crossings are the variants which display this type of 'crotch-defence' most clearly.

Turning from postures to movements, a great deal has been written about gait. The walking styles of different individuals and of dif-

■ *Seated leg postures show marked gender differences. Sitting with the legs apart or with the legs crossed ankle-on-knee, are typically male positions. Sitting with the legs twined so that the crossed-over foot rests against the side of the other leg is an exclusively female position. Clasping one tightly-bent leg is a genital-defence posture.*

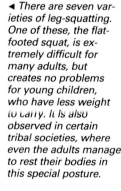

◀ *There are seven varieties of leg-squatting. One of these, the flat-footed squat, is extremely difficult for many adults, but creates no problems for young children, who have less weight to carry. It is also observed in certain tribal societies, where even the adults manage to rest their bodies in this special posture.*

ferent cultures have fascinated observers for many years. Personal differences are enormous and many famous people have such distinctive walking actions that they can be imitated with ease. One only has to mention the names of Charlie Chaplin, Groucho Marx, James Cagney, Mae West or John Wayne to illustrate this. At a cultural level there are huge differences between, say, the Japanese and the Americans. A detailed scientific analysis of these variations in gait remains to be done, but it is possible to provide a preliminary classification of them:

1. The Stroll. This is the ordinary, slow gait used when walking is an end in itself. It operates at a speed of about one step per second.

2. The Amble. This is even more relaxed than the stroll and may involve frequent, vague changes in direction and brief pauses. It is a meandering gait, often employed as a time-filler.

3. The Saunter. This is a slightly more rhythmic gait than the stroll, the performers being more 'on display' as they saunter up and down. It is a form of mobile loitering, aimless in direction but *publicly* demonstrating a leisurely mood.

4. The Dawdle. The gait that 'hangs back' and hesitates. The dawdler repeatedly slows down and frequently stops, as if not wishing to progress forward for some reason. It is a sluggish, lingering gait.

5. The Plod. This is a heavy-footed gait performed when the walker is tired or depressed. Each foot is brought down forcefully while the knees are kept slightly bent. Even those who are not tired adopt this type of progression when ascending a steep hill.

6. The Slouch. Another tired gait, but this time without the heavy footfall. Instead the body leans forward in a slightly hunched pos-

■ With our well-developed calf and thigh muscles we are able to walk, run, dance, kick, leap and swim with great vigour and for considerable periods of time. The record for non-stop walking is 357 miles in six days. The figure for non-stop running is remarkably similar: 353 miles in five days. For non-stop swimming it is slightly less: 299 miles, but this was achieved in three-and-a-half days.

ture. The tilt of the body helps the forward progression. This stooping gait is typical of subordinates who have accepted their subservient role in life.

7. The Waddle. A slow, swaying gait performed by the grossly overweight and by those with feet defects. The inefficiency of the walking motions causes the body to sway from side to side with each step.

8. The Hobble. The gait of those whose legs cannot perform full strides with comfort. The hobbler moves forward, usually with pain,

using very short steps. This is typical of people with damaged feet or those wearing very tight skirts or cramped shoes.

9. The Totter. This is an unsteady gait caused by physical frailty in the sick or aged, or by an excessive intake of alcohol at any age. The body attempts normal walking, but the balancing mechanism keeps failing and a few side-steps occur before the original course is regained.

10. The Limp. This is an asymmetrical gait caused by damage to one leg only. Although the limper may be progressing with some urgency, his forward movement remains slow because of his lameness.

11. The Shuffle. The foot-dragging gait of the sick and the seedy. This is the walk of the hospital patient following an operation, or of an elderly down-and-out on the streets.

12. The Prowl. The illicit gait of someone who does not wish to be detected. The body leans slightly forward and the feet are placed gently on the ground toes-first to reduce noise.

13. The Tip-toe. An extreme form of the prowl employed for maximum quietness, with the knees bent and only the toes touching the ground with each step.

14. The Promenade. This is halfway between the slow stroll and the full-speed walk. It could be called a fast stroll or a leisurely walk. It sets out to cover a certain route, but purely for the pleasure of the trip.

■ The human legs have great versatility, enabling us to perform many different types of bipedal locomotion. Some of these have been raised to the level of competitive events; others, such as the stylized mincing of the fashion model on the catwalk, have developed as cultural specializations which are limited to one specific social context. Most of the 36 listed types of human locomotion are however widespread and informal.

The 'releasing' power of rhythmic leg movement is exploited in tribal dance and ritual.

15. The Walk. Although everyone has a unique gait, it is possible to imagine a neutral, typical walk for the human species. It would be performed at a rate of roughly two steps per second, with the heel touching the ground first at each footfall. At any given moment either one or both feet would be touching the ground, which is the main difference between walking and running.

16. The Mince. To mince is to chop up into small pieces and a mincing gait is one in which fast but very short steps are taken. In effect, it is an exaggeration of the female walking style, the short steps of which are made even more abbreviated. It is a type of gait described as displaying 'affected preciseness'.

17. The Glide. This is an elegant version of the mince. By short, delicate movements of the feet, the body seems to glide forward as if on wheels. Once common among high-status females in parts of Europe, it is now confined largely to Japan. To create its impact it requires the wearing of a full-length skirt which hides the movements of the feet.

18. The Bounce. Teenagers and young adults often walk with a springy step that bounces the body with each pace. It is a joyous gait visibly demonstrating health and optimism.

19. The Strut. More jerky than the Bounce and with a stronger element of asserted self-confidence, the strut is typical of those who wish to impress the people near them.

20. The Swagger. Similar to the strut, but with added swaying movements. Even more boastful and self-satisfied. This gait has about it an air of confidence.

21. The Roll. A friendly, unthreatening version of the swagger. A slightly awkward walk in which the performer rolls from side to side as he or she progresses, emphasizing this action by bending towards people when pausing to talk to them.

22. The Stride. This is a cool but dominant gait characterized by unusually large paces. This is typical of high-status males and also, today, of females imitating the forcefulness of the more powerful male gait.

23. The Tramp. This is a vigorous version of the slower plod. It is a heavy-footed walk used on long treks and is typically observed in rough country where the feet have to be picked up more than usual, between each footfall.

24. The Lope. The walk of a less dominant individual whose slightly hunched posture and forward tilt gives the impression that if he stopped walking he would fall forward.

25. The Slink. The secretive gait of someone who is submissive and fearful. The slinker sidles along trying to be inconspicuous and to make his or her presence as unfelt as possible.

26. The Wiggle. The erotic walk of the female who wishes to display her gender signals to the maximum. The weight is placed first on one hip and then on the other. If over-emphasized, this gait quickly becomes a sexual joke.

27. The Dart. The anxious female gait, full of short, indecisive darting movements, this way and that, with much flutter and birdlike change of direction. Male darters appear fussy and effeminate.

28. The Slog. A fast, heavy-footed walk that is almost a march, but far less formal. Employed by men or women who are on the move in

■ *Rhythmic jumping as part of synchronized dancing has a powerful emotional impact on those who perform it repeatedly over a long spell of displaying. Dancers performing in this way seem to enter an almost trancelike state. The western celebrity was simply asked to jump in the air for the camera.*

■ *Leg decorations have included both painful tattooing and painless painting. In addition, some tribes have insisted on their females wearing heavy rings on their legs, or tight bindings, both of which have restricted their movements and destroyed their independence.*

difficult terrain and have to travel a long distance in a limited time. Slogging along, they give the impression of a controlled response to urgency.

29. The Hurry. A fast, light-footed walk that has about it an air of uncontrolled urgency. It is on the verge of breaking into a run, but just stops short of doing so. The step-rate remains the same as in the ordinary, unhurried walk, but the stride is lengthened slightly.

30. The Bustle. A fast, light-footed walk in which the performer keeps changing direction, scuttling here and there in a state of agitation. It is a less birdlike version of the dart.

31. The Prance. A playful fast walk in which unnecessary springs and small leaps are made with the feet as the person progresses forward. It is a fast version of the bounce, with a more vigorous leg action.

32. The Jog. A slow, controlled version of the run. The running action is performed at the slowest possible speed without destroying its rhythm. Nowadays it is frequently performed with earnest intent by the totally unfit and the no-longer-young, for whom it is more likely to be a health hazard than a health aid.

33. The March. The formalized military version of the fast walk, in which the stride is lengthened and the swinging of the arms is greatly increased. The arms are stiffened and their swinging helps to balance the body as it takes its longer, fast strides.

34. The Goose-step. A dramatically exaggerated form of marching, in which the legs are kicked forward with each pace and kept in a permanently stiff, straight posture, with knee-bending completely and artificially eliminated. Strictly for parades.

35. The Run. This was the gait so crucial to our primeval ancestors on the hunt. The body leans forward and the runner pushes hard

■ *To kneel before someone is to perform a submissive act of body-lowering. Long ago it was agreed that a dominant human being should only be offered a one-knee kneel and that the two-knee kneel should be reserved exclusively for the deity. Today the two-knee version is only seen in formal contexts, during prayer, or in informal contexts when a sportsman feels a sudden urge to thank heaven for his moment of triumph.*

against the ground with his feet. In walking there were two feet or one foot on the ground at any given moment; in running there is one foot or no feet touching the ground.

36. The Sprint. The fastest form of human leg progression, the sprint employs footfalls which involve only the front part of the feet, the heels hardly touching the ground at all. The stride rate increases to four or five per second, but this can only be maintained for very brief periods. In a primeval context it would have been especially valuable during the climactic moments of the hunt.

These are the main forms of locomotion and each of us employs most of them during our lifetime. Some are triggered by emotional conditions while some are the result of social rules. These rules have varied from epoch to epoch, and in more formal times strict conditions were laid down concerning the manner in which a gentleman should walk in the street. Here is an example from the early eighteenth century: 'A Gentleman ought not to run or walk too fast in the Streets, lest he be suspected to be going a Message; nor ought his pace to be too slow; nor must he take large Steps, nor too stiff and stately, nor lift his Legs too high, nor stamp hard on the Ground, neither must he swing his Arms backward and forward, nor must he carry his knees too close, nor must he go wagging his Breech, nor with his feet in a straight Line, but with the In-side of his Feet a little out, nor with his Eyes looking down, nor too much elevated, nor looking hither and thither, but with a sedate countenance.'

These rules of 'good deportment' sound bizarre to us today, when we simply walk out of the front door and down the street without giving a passing thought to the way we are placing one foot in front of another. For the leg-watcher, this new informality is a bonus because it has allowed personal gaits to develop unhindered by the restrictions of etiquette, and provides a much greater variety of walking actions.

Finally, there are a number of regional gestures involving the legs. Slapping the thigh has a wide variety of meanings in different parts of the world. In South America it signifies impatience or anger, the slap on the thigh being a mimed act of what the person would like to do to someone else. In Europe it is more likely to indicate surprise, shame or sorrow. In Ancient Greece and Rome it could mean horror or mourning, but was also used to demonstrate intense joy. The only common factor shared by all these messages is one of a sudden, sharp emotional response.

There are also various submissive knee gestures – the bending of the knee in the curtsey, the one-knee kneel and the two-knee kneel. These body-lowering devices have been common and widespread throughout history, but are comparatively rare today, the simple forward bow of the body having largely replaced them. The curtsey is largely limited to women greeting royalty; the one-knee kneel is restricted to males receiving knighthoods, and the two-knee kneel is now strictly an act of subordination to God during prayer. In earlier days, however, the curtsey was widely used as a polite greeting, and kneeling was offered to a whole range of dominant individuals in addition to royalty. In these days of greater equality, however, the bending of the leg as a genuinely submissive act has become a gesture of great rarity.

THE FEET

It has been said that man stands alone because he alone stands. To put it another way, the first great step for mankind was the first bipedal step taken by our remote ancestors. The moment we started walking on our hind legs we freed our front legs to become our grasping, manipulative hands. And with tool-making hands we conquered the world.

So we owe a great debt to our feet and should revere them as one of the most important parts of our anatomy. Perversely, we fail to do this. Instead we abuse them horribly. We sentence them to spend two-thirds of their life inside cramped leather cells. We force them to walk on hard, tiring surfaces, and we completely ignore their health and well-being until they are in serious trouble and sending out pain signals we can no longer ignore.

The reason we look down on them, metaphorically, is that we look down on them literally. They are too far away from our specialized sense organs. If we could examine them as closely as we study our hands, we would take more care of them; but they are at the far end of the body, and most of the time they hardly rate a passing thought.

This attitude is fostered by the feeling that damage to our feet cannot be lethal. Even when a woman kicks off her high-heeled shoes after a shopping trip and exclaims 'My feet are killing me!', she does not really believe it. But her statement contains more truth than she might realize. It is an undeniable fact that the foot is not a 'vital organ' like the heart, lungs or liver; but badly treated feet can shorten a lifespan just as surely as a heart attack. To understand why this should be it is necessary to make some field observations on the walking behaviour of the elderly. Those who have thoughtlessly abused their feet for decades are finding old age a time for hobbling painfully along at a snail's pace. Others who still have efficient walking feet can stride out for long 'constitutionals'. Taking long walks in old age has emerged as one of the best life-stretchers there is. A survey of exceptional individuals who live into their nineties and beyond reveals that a remarkably high proportion of them have been devoted walkers, often covering several miles a day, every day. There is something about relaxed walking that exercises the whole body in an ideal way. The currently fashionable jogging habit, on the other hand, can cause all kinds of problems, except for younger adults. The feet love gentle movement; they hate jarring movement.

Every time the foot touches the ground as we move forward, no matter how soft the step, it receives a jolt. It has been estimated that, in an average life of moderate activity, the feet hit the ground more than ten million times. The first moment of each contact consists of the heel pad slamming down and acting as a shock absorber. We take this vital action completely for granted, but it is only necessary to miss a step up or a step down in the dark to realize how unpleasant it is to touch the ground in some other, unprepared way.

The split second following this initial contact, the foot has switched roles. From being a shock absorber it now becomes a rigid support structure for the moving weight of the body. Finally, through its toes, it becomes a pushing organ projecting the body forward. This triple sequence occurs with every step we take.

To make all this possible, the foot is a remarkably complex structure. It contains 26 bones, 114 ligaments and 20 muscles. Leonardo

▼ The human foot, described by Leonardo da Vinci as 'a masterpiece of engineering and a work of art', comprises 26 bones, 114 ligaments and 20 muscles.

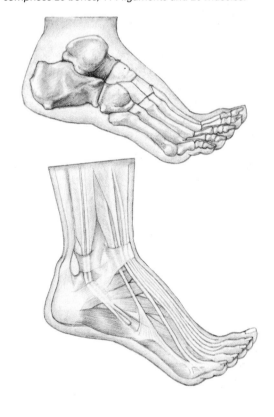

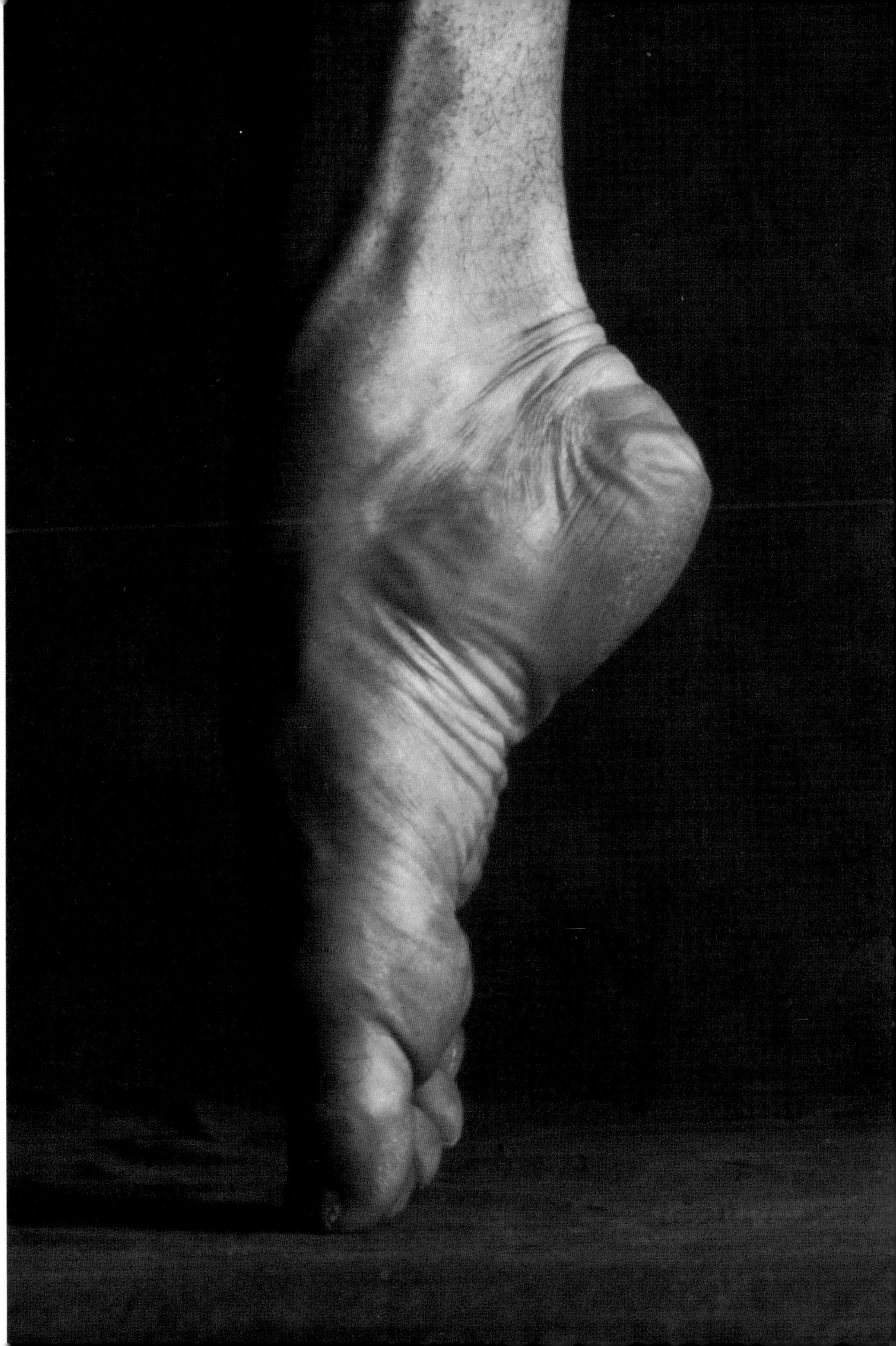

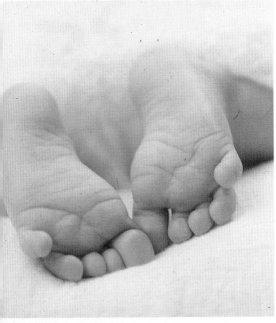

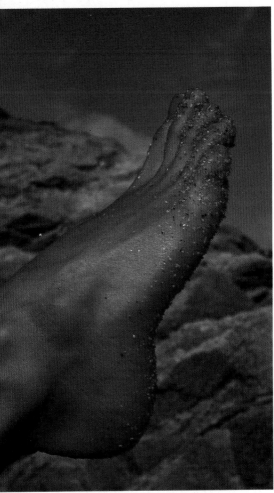

■ *The toes of a baby's foot are 20 times better at grasping objects than those of an adult human. A newborn foot is some three inches long; in females it about triples in length by the time it is fully grown; in males it grows to about three-and-a-half times its baby size. Although for most adults the toes remain rather weak and inactive parts of the human anatomy, for ballet dancers toe strength becomes a factor on which entire careers depend. The foot opposite is that of Rudolf Nureyev.*

da Vinci called it a 'masterpiece of engineering' – and when you consider the special kind of balancing trick it must perform for our uniquely upright bodies, you are forced to agree with him. Imagine, for instance, a solid, life-size dummy of a standing human being, with the weight distributed in a natural fashion, and consider what would happen if it were to be given a gentle push. It would crash over immediately, a top-heavy disaster. Imagine what would happen if such an object were placed on the side of a hill or on some kind of sloping ground. It would topple over instantly. Despite this we are remarkably nimble. This is because the feet are sending and receiving countless messages during every second of human movement, resulting in thousands of minor muscle adjustments enabling us to keep a balanced view of the world. Even when we are standing still and seemingly inactive, the feet are busily working away, making tiny, subtle and almost imperceptible alterations in our posture.

To achieve all this we have had to make one special sacrifice during the course of evolution. As one anatomist put it, rather colourfully, we had to develop 'webbed feet'. What he meant by this was that our big toes had to give up their opposability and become 'welded' to the other toes. In technical terms, this meant that the transverse metatarsal ligament had to spread across all five toes instead of only four. In apes, the metatarsal bones of the big toe are free from the rest of the foot and this makes for a much longer, more prehensile big toe. In humans the five toes are all shorter and more tightly 'webbed' together. We can still wiggle our toes, but we have lost our natural ability to grasp with them.

One ability we have not lost is leaving scent signals with our feet. it is claimed that Australian Aborigines have been able to tell the identity of individuals by smelling their footprints some time after they have passed by. In such cases, of course, the walkers in question were barefoot, but dogs can track human footprints even when the individuals concerned have been wearing thick shoes. This is possible because the soles of the feet are more richly supplied with sweat glands than any other part of the body except the palms of the hands (which are themselves ex-feet, of course). These sweat glands are highly susceptible to stress and they increase output dramatically whenever we are under pressure. We become aware of our 'sweaty palms', but we do not always realize that our feet are following suit. The scent produced is so strong that enough of it can leak out through our socks and our shoes to leave a scent trail that, even when two weeks old, is child's play for any bloodhound's nose. It seems highly probably that in our primeval, barefoot past, when our species was much thinner on the ground, our foot-scent signals were of some considerable importance to us in keeping tabs on both our friends and our foes.

Today the odour-producing ability of our feet has become nothing but a nuisance. Only toiletry manufacturers benefit by it. Because the scented sweat is trapped inside the prison of socks and shoes, it quickly falls prey to bacterial action and goes stale. Our marvellous scent-signalling feet have become a travesty of their earlier selves.

Another property of our feet that has largely lost its use is the ridging of the ventral skin. We have toe-prints that are every bit as individual as our finger-prints and which could be used in the same way to identify us. Their original, anti-slip function has become almost

Lucky the barefoot baby who can wiggle his toes.

▲ *The baby's feet grow fast, as these 'birth-prints' reveal when compared with the now much larger foot. But proud parents would do well not to rush the walking process. Infant feet often suffer from premature attempts to make the child take its first, wobbly steps. They are also damaged by cot sheets tucked in too tightly and by the enforced wearing of tiny 'baby-shoes'. Lucky is the barefoot baby who can wiggle his toes freely, as the feet mature and strengthen.*

meaningless in cultures where the wearing of shoes is the norm.

This ridged skin on the soles of the feet and the palms of the hand is called *volar skin* and it has one very strange property. It never gets suntanned. The obvious reply to such a statement is to point out that these are the two areas which are usually hidden from the sun, but this is not the true answer. If the palms and soles are deliberately exposed to the sun, they still remain untanned. Something in the human body specifically inhibits the production of additional melanin in these regions, leaving the palms and the soles a paler colour than the rest of the sun-tanned body. Even very dark-skinned races still retain pale soles and palms, suggesting that this quality is part of the evolutionary heritage of our whole species. The explanation put forward by those who have studied this phenomenon is that hand gestures and foot gestures are made more conspicuous by this device. With hand gestures this is easy to accept, but it may seem rather far-fetched to suggest that sole-of-the-foot gestures were ever that important. The examination of foot movements in moments of emotional conflict, which follows later, will make this less difficult to understand.

In tribal cultures where it is still the norm to go barefoot, it is possible to see the strength that can be developed in this lowest part of the human body. The Samoan, Fuatai Solo, climbing a coconut tree, has to be seen to be believed. He has been known to shin barefoot up a 30-foot tree-trunk in less than five seconds. By comparison all 'well-heeled' urbanites come into the category which western ranchers used to refer to as 'tenderfoot'.

Fuatai Solo set up his barefoot record in Fiji in 1980; and it is on the Fijian Islands that one can still observe an even more incredible achievement of the human sole: fire-walking. The event begins with a long period of prostrate relaxation. The fire-walkers, who will perform in the evening, gather and lie down quietly together for several hours. Then, after dark, they light logs of wood in the fire-pit and begin to heat the large smooth stones (huge pebbles collected from the shore) which lie packed tightly together beneath the burning branches. When the fire has reached such a heat that the pebbles are glowing, they start to rake away all the embers until only the stones are left, fully exposed and still hot enough to ignite a handkerchief if one is dropped on them. At this point, unbelievably, the fire-walkers, in their bare feet, walk nimbly across the baking stones as if crossing a shallow river on a group of stepping-stones.

Logic demands that the soles of their feet should be seriously blistered – that the 'meat' of their feet should be cooked – but they remain unharmed. I have personally examined the soles of the feet of these fire-walkers immediately after their performance and, to my surprise, found that they were unusually soft and spongy. They were certainly not calloused or horny, nor were they secretly treated in any way. I also examined the stones in the fire-pit and found that they were still intensely hot the *morning after* the fire-walk. I have no explanation for this amazing achievement of the human body. Other investigators have found themselves similarly baffled, and the explanations put forward seem grossly inadequate. The best attempt suggests that when the skin comes into contact with intensely hot surfaces the natural moisture of the body vaporizes so fast that it forms a protective layer between the skin and the stone. Although it is just possible to

▶ *Under normal circumstances the human foot is allowed little precision work, but with training and perseverance it is possible to write with the feet or even to shave a customer in a barber's chair. For those with upper-limb disabilities this potential can prove immensely rewarding.*

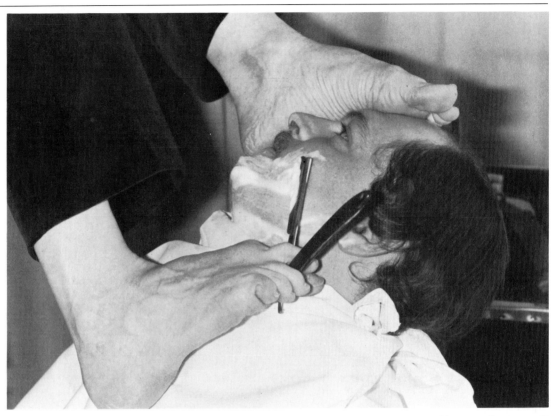

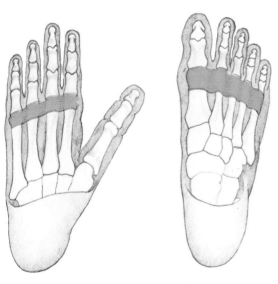

■ *The ape's foot (left) is a twin-function organ, being both an efficient grasper and a walker. The human foot, although capable of holding small objects under special circumstances, is essentially a locomotory organ, with the grasping gap between the big toe and the others reduced to a minimum. During evolution it has been converted from a generalized organ to a specialized organ.*

conceive of such a process, and to imagine a sort of human hovercraft walking along on a thin cushion of rapidly expanding vapour, the whole idea collapses when one recalls the last time one touched a hot stove, screamed with pain and suffered a severe blister. For the moment, the fire-walking ability of the human foot must remain a fascinating enigma.

When we are born our tiny feet are soft and floppy, and about one-third of the adult length. It takes them twenty years to complete their long, slow growth, and it is a mistake to rush them. Eager parents who try to persuade their offspring to walk before they are ready to do so may actually cause harm to their feet. Even more damaging are tightly restricting bedclothes. These may tuck the baby in snugly, but if the sheet is too taut over the lower legs, the softly pliable feet may be twisted and squashed while the infant sleeps. Stiff, constricting shoes and clinging socks can also compress the soft infant foot and all these impositions may, in extreme cases, stretch young ligaments out of shape and throw the soft bones out of alignment.

Further damage is done to the growing feet during schooldays, when parents delay replacing out-grown shoes. Tight shoes or boots crush the toes and cause permanent damage. For males, reaching adulthood solves this problem, but for many females a new hazard arrives: the high-heeled, pointed-toe fashion shoe. This enemy of the human foot has been with us for centuries and shows little sign of retreating, despite our knowledge of the trouble it causes. Even the fact that women are four times more likely to suffer from foot ailments than men has not dented its popularity.

The reason for women's addiction to unsuitable, uncomfortable shoes is not hard to find. Foot-size is a gender signal. Since the female

Feet that can walk on glowing coals.

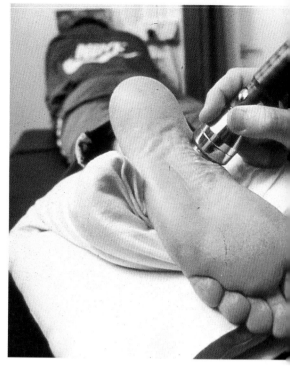

■ *The soles of the feet are soft and vulnerable to harsh surfaces. Professional sportsmen may spend long hours receiving expert massage of this region of the body. Despite this, some humans have the ability to perform firewalking on surfaces which should, in theory, blister – if not roast – the soles of the feet. No adequate theory has been put forward to explain the strange capacity. There are two variations on the fire-walking theme – the Fijian technique of walking across hot, smooth stones, and the Asiatic method of walking along a path covered with hot cinders. Some westerners, like the Harvard Medical School professor seen here, have successfully mastered this activity without injury or apparent pain, though how they manage this remains a mystery.*

foot is both shorter and narrower than its male counterpart, it naturally follows that a daintier foot is more feminine and therefore more attractive to the male. The sharp point of the typical female shoe has a streamlining effect and its high heel changes the posture of the foot in a way that makes it seem shorter.

Unfortunately this type of fashion shoe disturbs the balance of the female body, causing leg pains, back-ache and even headache, but the deep-seated fear of the ugliness of the big-footed body drives women on. Words like 'clodhopper' and 'goosefoot' do little to help, and the old jazz song 'I don't love ya 'cos your feet's too big', which includes the classic line 'your pedal extremities are obnoxious', re-inforces the message.

The passion for small female feet reached such intensity in earlier centuries that some ladies of fashion were known to have had their small toes amputated to give them an extra advantage, when slipping their feet into even-more-pointed footwear. Mention of amputation inevitably brings to mind the cruel story of Cinderella. Today's

The soles are always pale; they even resist suntanning.

▼ Like the palms of the hands, the soles of the feet never become suntanned. Even dark-skinned races lack heavy pigmentation in these regions and this makes their hand and foot movements unusually conspicuous.

Disneyfied version of it is harmless enough, but the original was bloody and savage. A prince was looking for a wife, but to satisfy his demand for femininity, she had to have very small feet. A tiny fur slipper was used to test prospective brides. Two sisters were desperate to be chosen. The elder one tried to force her foot into the slipper but it would not fit, so her mother told her to cut off her big toe, explaining that once she was married to the prince she would never need to walk again, so there was little to lose. The girl chopped off her big toe and squeezed her bleeding foot into the slipper, but as she rode away with the prince, he noticed the blood oozing from the slipper and staining her stockings. He returned her to her mother, who offered him the other daughter. This time the unfortunate girl had to have her heel cut down in size to squeeze into the slipper. Again, spurts of blood gave the game away and she too was rejected. Only then did the prince find Cinderella, whose tiny foot was a perfect fit and who became the blushing bride of the princely foot-fetishist.

The bizarre premise of this story – that a high-ranking male will find a tiny-footed female sexually acceptable regardless of her other qualities – seems to have been overlooked by modern audiences. This is because the modern version of Cinderella has converted the two sisters into *ugly* sisters, while Cinderella is always very pretty. But this is cheating. The prince only made one demand of his bride – that her foot should fit the tiny fur shoe (not glass, by the way, which was a mistranslation of *vair* for *verre*). To understand why he put such emphasis on the foot alone, it is necessary to know that this story originated in China, where foot-binding of young girls had been a common practice among families of high rank for centuries. There, the smallness of a girl's foot was the all-important mark of beauty.

Chinese foot-binding took the following form. When a girl was tiny she was allowed to run about freely, but before long, usually between the ages of six and eight, she was subjected to the agony of having her toes tied in to her sole. A bandage two inches wide and ten feet long was wrapped over the four small toes, bending them cruelly back on themselves. It was then wound tightly around the heel, pulling the bent toes and the heel closer together. The rest of the bandage was wrapped around and around to ensure that the foot could not be forced open again into a normal position. Only the big toe escaped this punishment and was left unbound.

Girls who cried were beaten. Despite the pain they were forced to walk on their crushed feet, in order to force the feet into accepting their new, buckled shape. Every two weeks, a new pair of shoes was put on, always one-tenth of an inch shorter than before. The aim, incredibly, was to reduce the length of the foot to one-third of its normal size – to the much-prized 'Three-inch Golden Lotus'.

By the time they were adult such girls were permanently crippled, unable to walk normally and strictly limited in the physical activities they could perform. This was the social bonus of the deformity. Not only did they have super-feminine smallness of foot, but they were also literally unable to stray from their husbands. In addition, they provided a permanent display of high status, since they clearly could do no manual labour of any kind. Only with the modernization of China in the present century and the sweeping away of Mandarin society, was this extraordinary form of mutilation stamped out.

Natural differences between male and female feet have sometimes been cruelly exaggerated.

One of the reasons for the appeal of the bound foot of the Chinese female was sexual. The Golden Lotus, as the tiny foot was called, had erotic significance in several strange ways. The girls' lovers were said to enjoy not merely kissing their feet during sexual foreplay but actually taking the whole foot into their mouth and sucking it avidly. The more sadistic lovers enjoyed the ease with which they could make their women scream during love-making simply by squeezing their crippled feet. Furthermore, by placing the two feet together, their buckled shape formed a pseudo-orifice which could be employed as a symbolic vagina. The real vagina was also said to be improved by the stilted form of walking caused by the bound feet: 'The smaller the woman's foot, the more wondrous become the folds of the vagina.' In addition to these and other, more outlandish erotic ideas about the Golden Lotus, there was a general sexual excitement in the idea of the helplessness of the females with bound feet. With their localized form of bondage they were at the mercy of their men, and suffered at their hands for centuries.

Leaving China, the general symbolism of the foot is often sexual elsewhere, even without the bondage factor. There has been a widely held belief that unusually large feet in a man mean that he has a large penis, while unusually small feet in a woman mean that she has a small vagina. But this is no more than a simplistic extension of the biological gender difference in foot size.

The shoe has frequently been employed as a symbol of the female genitals, and this is why 'The Old Woman who Lived in a Shoe' (in other words, whose life was centred on her genitals) 'Had so many children she didn't know what to do'. It also explains why shoes and boots are tied to the back of the cars of departing honeymooners and why a romantic lover used to drink champagne out of his lady's shoe. An old French tradition demands that the bride should keep her wedding shoes, and never give them away, if she wants to live happily ever after with her husband. And Sicilian girls seeking a husband always slept with a shoe under their pillows. These and many other such customs confirm the symbolic link between the shoe and sex.

In other contexts, shoes have had other symbolic values. In ancient times they stood for liberty, because slaves went barefoot. The removal of shoes in certain places of worship presumably arose as an act of humility in the presence of the deity, for whom the worshipper was a 'willing slave'.

The feet themselves, in ancient times, were often seen as the site of the human soul and it has been pointed out that in Greek legends lameness indicates some defect of the spirit or moral blemish. An even more ancient symbolism sees the feet as the rays of the sun, with the swastika symbol arising as a 'sun wheel with feet'.

Turning from foot symbolism to the body language of the feet, an interesting fact emerges. The feet are undeniably the most honest part of the whole human body. Small movements and posture-shifts of the feet tell us the truth about a person's mood. The reason for this is that we seldom think what we are doing with our feet. When we meet other people we concentrate on their faces and we know that *they* concentrate on ours, so we become proficient liars with our smiles and our frowns. We put on the face we want others to see. But as you travel down the body, away from the facial region, the body language

■ *In the Orient the foot has been the victim of rampant sexism. Although the male foot has often been developed as a martial-arts 'weapon', the female foot was, for centuries, crushed into an abnormally shortened condition in the quest for improved feminine beauty. Because female feet are naturally smaller than those of the male, it was argued that by increasing this difference it would be possible to exaggerate their appeal. Chinese girls of high social standing aimed to reduce the length of their feet to roughly one-third of the normal dimension. This involved tight binding and prolonged agony, but even so the custom was difficult to eradicate when China advanced into the twentieth century. The hidden function of this practice was the partial immobilizing of the Chinese female, rendering her utterly dependent on the male.*

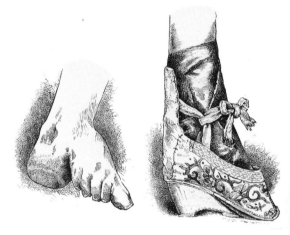

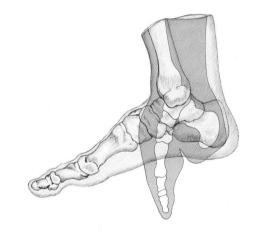

becomes progressively more sincere. Our hands are about halfway down and they are halfway honest. We are only vaguely aware of their actions, but we can lie with them to some extent. But the feet, at the other end of the body from the crucial facial region, are left to their own devices, and that is why they are so worthwhile studying.

The man who sits in his chair being interviewed looks so calm and relaxed. He is smiling softly and his shoulders are unhunched. His hands make smooth, gentle gestures. He appears completely at ease. But look now at his feet. They are wrapped tightly round one another as if clinging to each other for safety. Now he unwinds them and almost imperceptibly starts tapping one foot on the ground, as if he is trying to 'run away without moving'. Finally, he crosses his legs and the foot that now hangs in the air starts to flap up and down, again trying to flee while he stays on the spot. One famous interviewer in New York used to perform the 'foot flap' with such regularity that his television colleagues made a close study of it. They found that it only occurred when he was not at ease with his guest and they suggested that if he had the word HELP written on the sole of his shoe, the significance of his foot signal might become nationally understood.

Sometimes, the foot-tapping action of impatience, with its urgent signal of the desire to flee, becomes reduced to no more than a 'toe-wiggle', with the toes being raised and lowered almost invisibly. Like all foot-shifting and foot-jiggling movements these actions represent a suppressed desire to walk or run away from the situation being faced by the foot-mover. Lecturers, who would often like to flee from their audiences, perform a whole range of foot actions indica-

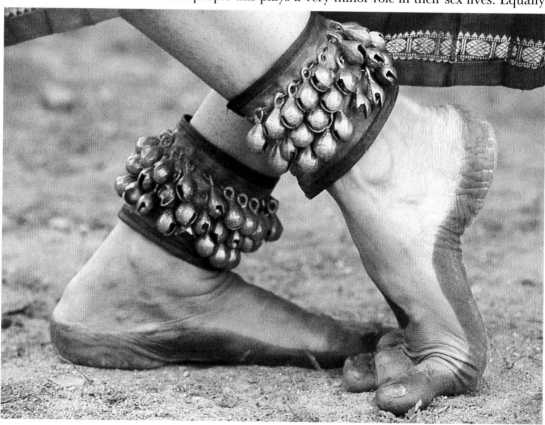

Langage des Jambes

J'hésite

Plus près

Ça Va!

De mieux en mieux

C'est formidable!

En plein ciel!

16

▲ Because the feet are so far away from our heads we tend to forget about them when we are deep in conversation. This makes them an unusually honest part of our anatomy, their unconscious actions revealing our true moods regardless of what our faces may be signalling. Only in certain courtship contexts do we start to communicate with deliberate, contrived foot-and-leg actions.

▶ The feet of the Indian dancer are often decorated with red patches of henna and the ankles adorned with small bells. This makes the stamping dance movements more conspicuous, both to the eye and to the ear. It is perhaps surprising that so few dancers in other parts of the world have used these adornment techniques.

ting their true mood. At a long conference, a study of foot movements of speakers is often more interesting than what they are saying. Unfortunately, it has become common practice for conference organizers to hide their speakers' bodies behind a lectern or some other barrier that prevents direct observation of their truthful lower appendages. Lecturer specialities, when they can be seen, include a delightful range of heel-raisings, foot-teeterings, swayings, pacings and tappings, as if the feet were trying every possible way of making their escape from the hostile stare of hundreds of pairs of eyes.

Boredom in a seated person with crossed legs is often expressed with yet another foot action – the mid-air multi-kick. This has a slightly more hostile tone to it than the foot-flap, as if the performer wishes to kick whoever is causing the boredom. The crossed-over leg repeatedly kicks forward into the air, but travels only a very short distance before returning to its original position. This is only a short step away from the full foot-stamp of anger.

Foot-shuffling and shoe-scuffing belong to a slightly different mood, typified by the small boy who is being quizzed about some misdeed. In his case, the movements of the feet lack the urgent, rhythmic beat of the multi-kicker or the foot-tapper. The boy's feet are not signalling outright attack or flight, or even marching defiantly off into the distance. Instead their irregular twists and shifts show that what they really want to do is to quietly sneak away.

Interpersonal contacts involving feet are few and far between, except for professionals such as chiropodists and masseurs. Lovers exploring each other's bodies may kiss each other's feet and toes, but for most people this plays a very minor role in their sex lives. Equally

▲ In the realm of sexual symbolism the shoe has nearly always stood for the female genitals. This is why the old woman who lived in a shoe, in the popular nursery rhyme, 'had so many children she didn't know what to do.'

► When the Pope wishes to perform an act of Christian humility to counteract the pomp of his high church station, he sometimes chooses the subordinate act of washing and kissing the feet of social inferiors. This involves the servile act of body-lowering before the other individuals and is closely related to his now well-known gesture of kissing the tarmac whenever he lands at a foreign airport.

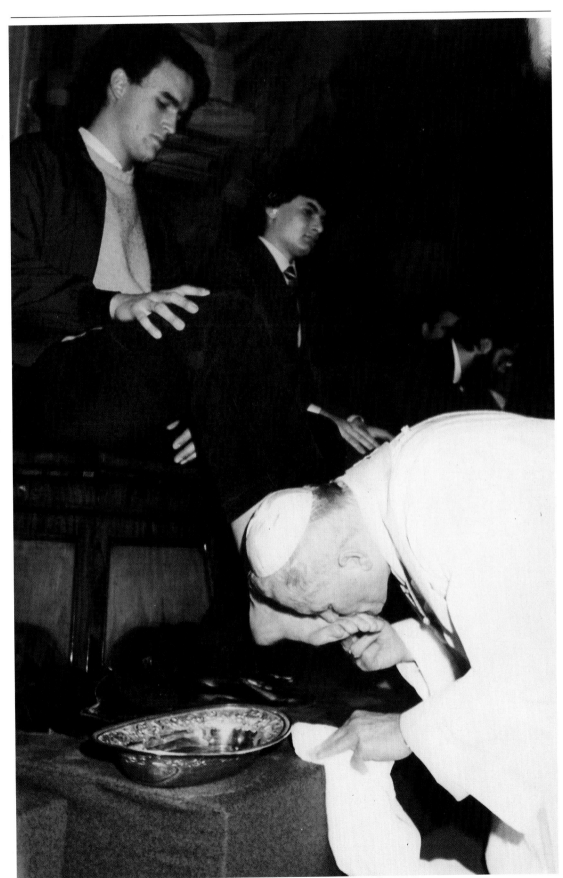

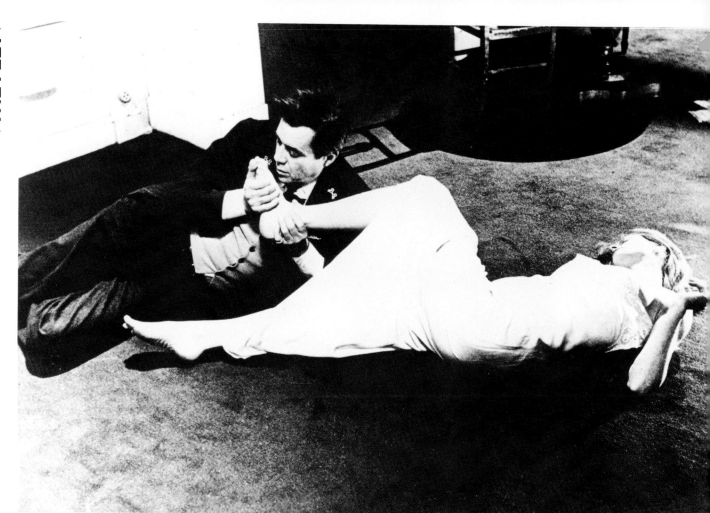

rare today is that other form of foot-kiss – the humble act of lowering the body to kiss the lowest part of a dominant figure. This is an act of extreme submission and subordination of a kind that is alien to modern societies. In ancient times, when rulers enjoyed a loftier status, the lips of the lowly and the feet of the lordly met on many an occasion. Diocletian, the Roman emperor who ruled as an absolute monarch, insisted that senators and other dignitaries kiss his foot when he received them and again when they left his presence. Even relatives of Roman emperors had to kiss the Imperial Foot. Today such activities are common only among foot fetishists, who pay specialized prostitutes large sums of money to be allowed to grovel at their feet.

The opposite side of the coin from grovelling at someone's feet is placing your foot on top of someone. This is an act of dominance which has been formalized in a number of contexts, the most familiar of which is the brave hunter who, having gunned down some innocent wild beast, stands proudly by his kill with gun in hand and one foot planted firmly on the back of the corpse. An old Polish Jewish custom involved treading on the foot of the spouse at the wedding ceremony. Whichever partner managed to do this first was supposedly destined to become the dominant member of the marriage. For 'upper foot', read 'upper hand'.

Although many of these old customs are dying out today, in Scotland the ceremony of the First Foot still survives. In this, good fortune for a coming year depends on the arrival on your doorstep of the First Foot of the New Year, no more than a few minutes after midnight as December 31st becomes January 1st. The newcomer must be carrying gifts, and, if the household is to prosper for the next twelve months, he must be a dark male stranger who does not suffer from flat feet. it is vital that he enters the house with his right foot, as the left foot is considered extremely unlucky. In earlier days, all persons entering a house had to ensure that it was their right foot which was placed first over the threshold. Grand residences employed special servants whose task was to ensure that nobody forgot this procedure. The reason for this preoccupation with left and right feet was that, as with other left and right distinctions, God was thought to operate through the right foot and the Devil through the left foot. To 'put one's best foot forward' was to stride out with the right foot. The right foot was good and kind; the left foot was wicked and hostile. This incidentally is why armies generally put the left foot forward first when setting off on a march. A typical order is 'Quick march! Left, right, left, right....' The hostile left foot is deliberately moved first to show the hostile intent of the marching men, although it is doubtful whether many modern-day soldiers are aware of this small piece of superstitious military history.

■ *Foot-kissing is not confined to Papal acts of humility. It is also a special form of erotic stimulation. There appear to be two reasons for this, one chemical and one masochistic. Chemically, the feet produce sweat rich in the same fatty-acids that are found in the genital region, so the foot-kisser may be experiencing unconscious stimulation of a primeval, olfactory kind. Second, the act of placing the mouth to the foot involves a 'grovelling' posture of servility which is exciting to masochistic personalities. As a special variant of this, some men find it sexually stimulating to be trampled on by women wearing spiky, high-heeled shoes.*

BIBLIOGRAPHY

Ableman, P. 1969. *The Mouth and Oral Sex*. Running Man Press, London.
Allen, M.R. 1967. *Male Cults and Secret Initiations in Melanesia*. Melbourne University Press, Melbourne.
Ambrose, J.A. 1960. *The Smiling and Related Responses in Early Human Infancy*. University of London. PhD Thesis.
Amphlett, H. 1974. *Hats: A History of Fashion in Headwear*. Sadler, Chalfont St Giles, Buckinghamshire.
Angeloglou, M. 1970. *A History of Make-up*. Macmillan, London.
Argyle, M. 1967. *The Psychology of Interpersonal Behaviour*. Penguin, Harmondsworth, Middlesex. — 1969. *Social Interaction*. Methuen, London. — (Editor) 1973. *Social Encounters. Readings in Social Interaction*. Penguin Books, Harmondsworth, Middlesex. — 1975. *Bodily Communication*. Methuen, London. — and Cook M. 1976. *Gaze and Mutual Gaze*. Cambridge University Press, Cambridge. **Austin**, G. 1806. *Chironomia; or, a treatise on rhetorical delivery*. London. **Ayalah**, D. and I.J. Weinstock. 1979. *Breasts*. Hutchinson, London.

Barakat, R.A. 1973. 'Arabic gestures.' *J. Popular Culture* pp749-787. **Barnard**, C. 1981. *The Body Machine*. Hamlyn, London. **Barsley**, M. 1966. *The Left-handed Book*. Souvenir Press, London. — 1970. *Left-handed Man in a Right-handed World*. Pitman, London. **Bauml**, B.J. and F.H. Bauml. 1975. *A Dictionary of Gestures*. Scarecrow Press, Metuchen, N.J. **Beck**, S.B. 1979. 'Women's somatic preferences'. IN: *Love and Attraction* (eds: Cook, M and G. Wilson). Pergamon, Oxford. **Bell**, C. 1806. *Essays on the Anatomy and Philosophy of Expression*. London. — 1833. *The Hand its Mechanism and vital Endowments as invincing Design*. Pickering, London. **Benthall**, J. and T. Polhemus. 1975. *The Body as a Medium of Expression*. Allen Lane, London. **Berg**, C. 1951. *The Unconscious Significance of Hair*. Allen and Unwin, London. **Bettelheim**, B. 1955. *Symbolic Wounds*. Thames and Hudson, London. **Birdwhistell**, R.L. 1952. *Introduction to Kinesics*. University of Louisville, Kentucky. **Broby-Johansen**, R. 1968. *Body and Clothes*. Faber and Faber, London. **Brooks**, J.E. 1953. *The Mighty Leaf: the Story of Tobacco*. Alvin Redman, London. **Brophy** J. 1945. *The Human Face*. Harrap, London. — 1962 *The Human Face Reconsidered*. Harrap, London. **Brownmiller**, S. 1984. *Femininity*. Hamish Hamilton, London. **Brun**, T. 1969. *The International Dictionary of Sign Language*.

Wolfe, London. **Bulwer**, J. 1644. *Chirologia; or the Naturall Language of the Hand. Whereunto is added Chironomia: or, the Art of Manual Rhetoricke*. London. — 1648. *Philocophus; or the Deafe and Dumbe Man's Friend*. London. — 1649. *Pathomyotomia, or a Dissection of the Significative Muscles of the Affections of the Minde*. London. — 1650. *Anthropometamorphosis; Man Transform'd; or the Artifical Changeling*. London. (Re-issued in 1654 as: *A View of the People of the Whole World*.) **Burr**, T. 1965. *Bisba*. Hercules, Trenton, New Jersey.

Campbell, B. 1967. *Human Evolution*. Heinemann, London. **Cannon**, W.B. 1929. *Bodily Changes in Pain, Hunger, Fear and Rage*. Appleton-Century, New York. **Chan**, P. 1981. *Ear Acupressure*. Thorsons, Wellingborough, Northants. **Chetwynd**, T. 1982. *A Dictionary of Symbols*. Granada, London. **Cohen**, H. 1979. *The Complete Encyclopedia of Exercises*. Paddington Press, London. **Coleman**, V. and M. Coleman. 1981. *Face Values*. Pan Books. London. **Comfort**, A. 1972. *The Joy of Sex*. Crown, New York. **Coon**, C.S. 1966. *The Living Races of Man*. Cape, London. **Cooper**, J.C. 1978. *An Illustrated Encyclopedia of Traditional Symbols*. Thames and Hudson, London. **Cooper**, W. 1971. *Hair: Sex, Society, Symbolism*. Aldus Books, London. — 1967. *Fashions in Eyeglasses*. Peter Owen, London. **Corti**, C. 1931. *A History of Smoking*. Harrap, London. **Coss**, R. 1965. *Mood-provoking Visual Stimuli*. University of California. **Critchley**, M. 1939. *The Language of Gesture*. Arnold, London. — 1975. *Silent Language*. Butterworths, London.

Darwin, C. 1872. *The Expression of the Emotions in Man and Animals*. John Murray, London. **D'Angelou**, L. 1969. *How to be an Italian*. Price, Sterne, Sloane. Los Angeles. **Devine**, E. 1982. *Appearances. A Complete Guide to Cosmetic Surgery*. Piatkus, Loughton, Essex. **Dickinson**, R.L. 1949. *Human Sex Anatomy*. Williams and Wilkins, Baltimore. **Dingwall**, E.J. 1931. *The Girdle of Chastity*. Routledge, London.

Ebenstein, H. 1953. *Pierced Hearts and True Love. The History of Tattooing*. Verschoyle, London. **Ebin**, V. 1979. *The Body Decorated*. Thames and Hudson, London. **Eden**, J. 1978. *The Eye Book*. David and Charles, Newton Abbott, Devon. **Efron**, D. 1972 *Gesture, Race and Culture*. Mouton, The Hague. (First published in 1941 as: *Gesture and Environment*. King's Crown Press,

New York.) **Eibl-Eibesfeldt**, I. 1970. *Ethology, The Biology of Behavior*. Holt, Rinehart and Winston, New York. — 1972. *Love and Hate. The Natural History of Behavior Patterns*. Holt, Rinehart and Winston, New York. **Ekman**, P. 1967. 'Origins, usage and coding of nonverbal behavior.' Centro de Investigationes Sociales, Buenos Aires, Argentina. — 1969. 'The repertoire of nonverbal behavior.' *Semiotica* 1 (1), pp49-98. — 1970. 'Universal facial expressions of emotion.' *California Mental Health Research Digest* 8 (4), pp151-158. — 1971. 'Universal and cultural differences in facial expressions of emotion.' *Nebraska Symp. on Motivation 1972*. — 1973. *Darwin and Facial Expression*. Academic Press, New York. — 1976 'Movements with precise meanings.' *J. Communication* 26 (3), pp14-26. — 1980. *The Face of Man*. Garland, New York. — et al. 1972. 'Facial expressions of emotion while watching televised violence and predictors of subsequent action.' *Television and Social Behavior* 5. pp22-58. U.S. Government Printing Office, Washington D.C. — and W.V. Friesen. (N.D.) 'Constants across cultures in the face and emotion.' *J. Personality and Soc. Psych*. — and W.V. Friesen. 1968. 'Nonverbal behavior in psychotherapy research.' *Research in Psychotherapy* 3. pp179-216. — and W.V. Friesen. 1969. 'The repertoire of nonverbal behavior: categories, origins, usage and coding.' *Semiotica* 1 (1), pp49-98. — and W.V. Friesen. 1969. 'Nonverbal leakage and clues to deception.' *Psychiatry* 31 (1), pp88-106. — and W.V. Friesen. 1972. 'Hand movements.' *J. Communication* 22, pp353-374. — and W.V. Friesen. 1974. 'Nonverbal behavior and psychopathology.' IN: *The Psychology of Depression: Contemporary Theory and Research*. Winston and Sons, Washington, D.C. pp203-232. — and W.V. Friesen. 1974. 'Detecting deception from the body or face.' J. Personality and Soc. Psych. 29 (3), pp288-298. — and W.V. Friesen. 1975. *Unmasking the Face*. Prentice-Hall, New Jersey. — and W.V. Friesen. 1976. 'Measuring facial movement.' *Envir. Psychol. and Nonverbal Behav*. 1 (1), pp56-75. — W.V. Friesen and P. Ellsworth. 1972. *Emotion in the Human Face*. Pergamon Press, New York. — W.V. Friesen and K.R. Scherer. 1976. 'Body movement and voice pitch in deceptive interaction.' *Semiotica* 16:1, pp23-27. — W.V. Friesen and S.S. Tomkins. 1971. 'Facial affect scoring technique: a first validity study.' *Semiotica* 3 (1) pp37-58. **Elias**, N. 1978. *The History*

of Manners. Blackwell, Oxford. **Elworthy**, F.T. 1895. *The Evil Eye*. John Murray, London.

Fast, J. 1970. *Body Language*. Evans, New York. **Fisher**, J. 1979. *Body Magic*. Hodder and Stoughton, London. **Fisher**, R.B. 1983. *A Dictionary of Body Chemistry*. Granada, London. **Fryer**, P. *Mrs Grundy. Studies in English Prudery*. Dobson, London.

Gabor, M. 1972. *The Pin-up, a Modest History*. Pan, London. **Gardiner**, L.E. 1959. *Faces, Figures and Feelings. A Cosmetic Plastic Surgeon Speaks*. Burstock Courtenay Press, Brighton. **Garfield**, S. 1971. *Teeth, teeth, teeth*. Arlington Books, London. **Gettings**, F. 1965. *The Book of the Hand*. Hamlyn, London. **Givens**, D.B. 1983. *Love Signals*. Crown, New York. **Glynn**, P. 1982. *Skin to Skin*. Allen and Unwin, London. **Gomez**, J. 1967. *A Dictionary of Symptoms*. Centaur Press, London. **Grigson**, G. 1976. *The Goddess of Love*. Constable, London. **Guletz**, S. (no date) *Hula!* South Sea Sales, Honolulu, Hawaii. **Guthrie**, R.D. 1976. *Body Hot Spots*. Van Nostrand Reinhold, New York.

Hall, E.T. 1959. *The Silent Language*. Doubleday, New York. **Harrison**, G.A. et al. 1964. *Human Biology*. Clarendon Press, Oxford. **Hendrickson**, R. 1976. *The Great American Chewing Gum Book*. Chilton Books, Radnor, Pennsylvania. **Henley**, N.M. 1977. *Body Politics*. Prentice-Hall, New Jersey. **Hennessy**, V. 1978. *In the Gutter*. Quartet Books, London. **Hess**, E. 1975. *The Tell-tale Eye*. New York. **Hess**, T.B. and L. Nochlin. 1973. *Woman as Sex Object*. Allen Lane, London. **Hewes**, G. 1983. 'The communication function of palmar pigmentation in man.' *J. Human Evolution* 12, pp297-303. **Hirschfield**, M. 1940. *Sexual Pathology*. Emerson Books, New York. **Hobin**, T. 1982. *Belly Dancing for Health and Relaxation*. Duckworth, London. **Hopson**, J.L. 1979. *Scent Signals*. Morrow, New York. **Huber**, E. 1931. *Evolution of Facial Musculature and Facial Expression*. The Johns Hopkins Press, Baltimore.

Inglis, B. 1978. *The Book of the Back*. Ebury Press, London. **Izard**, C.E. 1971. *The Face of Emotion*. Meredith, New York.

Jenkins, C. 1980. *A Woman Looks at Men's Bums*. Piatkus, Loughton, Essex.

Keogh, B. and S. Ebbs. 1984. *Normal Surface Anatomy*. Heinemann,

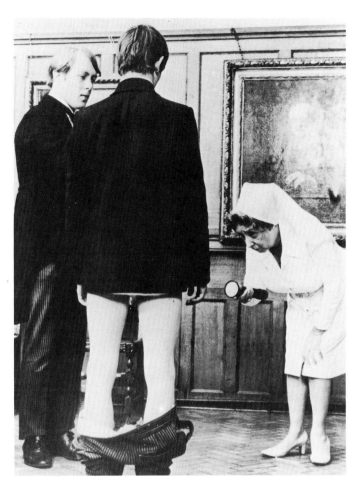

London. **Kinsey**, A.C. et al. 1948. *Sexual Behaviour in the Human Male.* Saunders, Philadelphia. — et at. 1953. *Sexual Behaviour in the Human Female.* Saunders, Philadelphia. **Kunzle**, D. 1973. 'The corset as erotic alchemy.' IN: *Woman as a Sex Object.* (Eds. Hess, T.B. and L. Nochlin.) Allen Lane, London.

Lamb, W. 1965. *Posture and Gesture.* Duckworth, London. **Lang**, T. 1971. *The Difference Between a Man and a Woman.* Michael Joseph, London. **Lavater**, J.C. 1789. *Essays on Physiognomy.* John Murray, London. **Lawther**, G. 1981. *The Healthy Body, A Maintenance Manual.* Muller, London. **Lee**, L. and J. Charlton. 1980. *The Hand Book.* Prentice Hall, Englewood Cliffs, New Jersey. **Lenihan**, J. 1974. *Human Engineering. The Body Re-examined.* Weidenfeld and Nicolson, London. **Levy**, H.S. (no date). *Chinese Footbinding.* Spearman, London. **Levy**, M. 1962. *The Moons of Paradise.* Arthur Barker, London. **Liggett**, J. 1974. *The Human Face.* Constable, London. **Lockhart**, R.D. 1979. *Living Anatomy.* Faber and Faber, London. **Lurie**, A. 1981. *The Language of Clothes.* Random House, New York.

Macintyre, M. 1981. *The Shogun Inheritance.* Collins, London. **Maclay**, G. and H. Knipe. 1972. *The Dominant Man.* Delacorte Press, New York. **Malinowski**, B. 1929. *The Sexual Life of Savages.* Routlege and Kegan Paul, London. **Mallery**, G. 1891. 'Greeting by gesture'. *Pop. Sci. Monthly*, Feb & March. **Maloney**, C. 1976. *The Evil Eye.* Columbia University Press, New York. **Mann**, I. and A. Pirie. 1946. *The Science of Seeing.* Penguin Books, Harmondsworth, Middlesex. **Mantegazza**, P. 1904. *Physiognomy and Expression.* Scott, London. **Mar**, T.T. *Face Reading.* Dodd, Mead, New York. **Masters**, W.H. and V.E. Johnson. 1966. *Human Sexual Response.* Churchill, London. **McGarey**, W.A. 1974. *Acupuncture and Body Energies.* Gabriel Press, Phoenix, Arizona. **Meerloo**, J.A.M. 1971. *Intuition and the Evil Eye.* Servire, Wassenaar. **Meredith**, B. 1977. *Vogue Body and Beauty Book.* Allen Lane, London. **Mitchell**, M.E. 1968. *How to Read the Language of the Face.* Macmillan, New York. **Morris**, D. 1967. *The Naked Ape.* Cape; London. — 1969. *The Human Zoo.* Cape, London. — 1970. *Patterns of Reproductive Behaviour.* Cape, London. — 1971. *Intimate Behaviour.* Cape, London. — 1977. *Man-watching.* Cape, London. — et al. 1979. *Gestures.* Cape, London. — 1981. *The Soccer Tribe.* Cape, London.

Munari, B. 1963. *Supplemento al Dizionario Italiano.* Muggiani, Milan.

Napier, J. 1980. *Hands.* Allen and Unwin, London. **Neumann**, E. 1955. *The Great Mother.* Routledge and Kegan Paul, London. **Nicholson**, B. 1984. 'Does kissing aid human bonding by semio-chemical addiction?' *Brit. J. Dermatology* III, pp623-7.

Papas, W. 1972. *Instant Greek.* Papas, Athens. **Parry**, A. 1971 *Tattoo. Secrets of a Strange Art.* Collier Books, New York. **Pease**, A. 1981 *Body Language.* Sheldon Press, London. **Penry**, J. 1971. *Looking at Faces and Remembering Them.* Elel, London. **Perella**, N.J. 1969. *The Kiss Sacred and Profane.* University of California Press. **Polhemus**, T. 1978. *Social Aspects of the Human Body.* Penguin Books, Harmondsworth. **Polhemus**, T and L. Proctor. 1978. *Fashion and Anti-fashion.* Thames and Hudson, London.

Reyburn, W. 1971. *Bust-up.* Macdonald, London. **Reynolds**, R. 1950. *Beards.* Allen and Unwin, London. **Rosebury**, T. 1969. *Life on Man.* Secker and Warburg, London. **Rudofsky**, B. 1974. *The Unfashionable Human Body.* Doubleday, New York.

Saitz, R.L. and E.C. Cervenka. 1972. *Handbook of gestures: Colombia and the United States.* Mouton, The Hague. **Scheflen**, A.E. 1972. *Body Language and the Social Order.* Prentice-Hall, Englewood Cliffs, N.J. **Scheinfeld**, A. 1947. *Women and Men.* Chatto and Windus, London. **Sheldon**, W.H. 1954. *Atlas of Men.* Gramercy, New York. **Shen**, P. 1982. *Face Fortunes.* Perigree Books, New York. **Sherzer**, J. 1972?. 'The pointed lip gesture among the San Blas Cuna.' IN: *Language and Society.* **Smith**, A. 1968. *The Body.* Allen and Unwin, London. **Sorell**, W. 1967. *The Story of the Human Hand.* Weidenfeld and Nicolson, London.

Taylor, A. 1956. 'The Shanghai gesture.' *F.F. Communications* No. 166. pp1–76. **Taylor**, R. 1970. *Noise.* Penguin, Harmondsworth, Middlesex. **Taylor**, W.P. 1983. *Bald is Beautiful.* Macmillan, London. **Thompson**, P. and P. Davenport. 1980. *The Dictionary of Visual Language.* Bergstrom and Boyle, London. **Tosches**, N. 1981. *Rear View.* Delilah Books, New York.

Ucko, P. 1968. *Anthropomorphic Figurines.* Szmidla, London.

Walker, B. *Body Magic.* Routledge and Kegan Paul, London. **Walls**,

G.L. 1967. *The Vertebrate Eye.* Hafner, New York. **Whiteside**, R.L. 1974. *Face Language.* Fell, New York. **Wildeblood**, J. 1973. *The Polite World.* David-Poynter, London. **Williams**, N. 1957. *Powder and Paint.* Longmans, London. **Wilson**, G. and D. Nias. 1976. *Love's Mysteries.* Open Books, London. **Winter**, R. 1976. *The Smell Book.* Lippincott, New York. **Woodforde**, J. 1968. *The Strange Story of False Teeth.* Routledge and Kegan Paul, London. — 1971. *The Strange Story of False hair.* Routledge and Kegan Paul, London. **Wood-Jones**, F. 1929. *Man's Place Among the Mammals.* Edward Arnold, London.

THE CREDITS

Lynda Poley
Picture research

Linda Proud
Picture-research coordination,
picture list and index

Michael Desebrock
Editorial coordination, picture
selection and design

Clive Sparling
Production

ARTISTS

David Mazierski
Drawings on pages 165, 216, 217

Andrea Newman
Drawings on pages 77,(upper), 96,
117, 140, 145, 173, 197, 221(lower),
237, 241

Francis Scholes
Drawings on pages 21, 24, 37, 49, 53,
65, 70, 77(lower), 80, 81, 100, 104,
205, 221(upper), 244

KEY TO PHOTOGRAPHERS

The key is in alphabetical order of surname.
In the picture list, the key-initials of
photographers are enclosed in brackets.

Ab	Abbas
dA	A. de Andrade
DA	Daniel Angeli
DAk	Danny Allmark
EA	Eve Arnold
J–LA	Jean–Louis Atlan
NA	Nitsuo Ambe
RA	Robert Azzi
AB	Arnaud Borrel
BB	Bruno Barbey
CB	Cecil Beaton
DB	Des Bartlett
GB	G. Bevilacqua
GeB	George Butler
GvB	Gert von Bassewitz
GBo	Gio Barto
HB	H. Benson
IB	Ian Berry
JB	Jane Bown
J/DB	Jan and Des Bartlett
MB	Mark Boulton
MBe	Marcus Brooke
MBs	Michael Boys
MBu	Michael Busselle
NB	N.Brown
RB	René Burri
RBd	R. Bond
RBr	Renaud Bachoffner
WB	Werner Bischof
WBn	W. Behnken
AC	Angelo Cozzi
ACy	Anthony Crickmay
BC	Bill Carter
C	Ciccione
Chim	David Seymour
GC G.	Chapman
HCB	Henri Cartier–Bresson
JC	Jan Cobb
LeC	LeCuziot
PC	Peter Carmichael
RC	Robert Capa
AD	Alan Davidson
BD	Bryan Duff
D	Delano
FD	François Duhamel
JD	John Doidge
JPD	Jean–Paul Dousset
RD	Raymond Depardon

SD	Shabbir Dossaji
SDe	Sergio Duarte
TD	Tony Duffy
EE	Elliott Erwitt
SE	Sarah Errington
F	Fedorenko
LF	Leonard Freed
L–F	Louis–Frederic
MF	Martine Franck
MFl	Michael Friedel
PF	Pepita Fairfax
PFo	Paul Fusco
RPF	Richard Phelps Frieman
AG	Alan Grisbrook, Cheadle
AGa	Ashvin Gatha
BG	Ben Gibson
BGg	Bob Gelberg
BGl	Burt Glinn
CvG	Cees van Gelderen, Amsterdam
EG	Ekkheart Gurlitt
FG	Felix Greene
HG	Helmut Gritscher
JG	Jean Gaumy
JGl	Jenny Goodall
LDG	Larry Dale Gordon
PG	Philip Gottop
RG	Ray Green, Manchester
RSG	Richard and Sally Greenhill, London
TG	Tim Graham, London
H	Hatani
DH	David Hurn
DWH	David W. Hamilton
EH	Ernst Haas
FH	Frank Herdholt
GH	G. Harrison
LH	L. Hestenberger
PH	Philippe Halsman
SH/JK	Stephane Husain/John Kacere
TH	Thomas Höpker
RJ	Rob Judges, Oxford
J	Jürgen
GK	G. Konig
KK	Kaku Kurita
MK	Mark Kaufman
RK	Richard Kalvar
WK	W. Klein
AL	Andrew Lawson, Charlbury, Oxon
Aln	Andy Levin
Alz	Annie Leibovitz
BL	Barry Lewis
BLt	Bertrand Laforet
GL	Guy Leygnac
HL	Hank Londoner
IL	I. Alex Langley
RL	R. Ian Lloyd
ReL	René Leveque
RoL	Robin Laurance
SL	Sam Levin
Sla	Sandra Lousada
AM	Alain Mingham
BM	Benn Mitchell
BMn	Butch Martin
CM	Constantine Manos
DM	David Montgomery
EM	Erling Mandelmann
GM	Ghislaine Morel
LM	Laurent Maous
MM	Michael MacIntyre
PM	Peter Marlow
PMd	Pierre Michaud
RM	Robert Mapplethorpe, NY
RMc	Robert McFarlane
AN	Alain–Patrick Neyrat
ANt	Albane Navizet
BN	Bob Nardell
JN	J. Nance
MN	Marvin Newman
RN	Robert Nicod
O	Obemski
MO	Margaret Olah
AP	Adrian Paul
CSP	Chris Steele–Perkins

GP	Gilles Peress
JP	Jesco von Puttkamer
JEP	J.E. Pasquier
JPP	J.P. Paireault
PP	Pierre Perrin
SP	Steve Powell
AR	Alon Reininger
DR	David Rubinger
JR	J. Reditt
LR	Leni Riefenstahl, Munich
MR	Mervyn Rees
GQ	Guy Le Querrec
S	Snowdon, London
AS	Art Seitz
BS	Brian Seed
DS	Dennis Stock
ES	Eric Schwab
HS	Heinz Stucke
HSs	Homer Sykes, London
JS	John Swannell
JSk	Jan Saudek, Prague
JSv	Jan Svab
RSf	R. St.Frank
SdeS	Serge de Sazo
SS	Sabastiao Salgardo Jr.
US	Ulli Seer
T	Tatiner
Tr	Trippett
PT	Pete Turner
ST	Steve Templeman
BU	Burk Uzzle
HU	H.R. Uthoff
AV	Avakadeo Vergani
FV	F. Vuich
JV	J.P. Vidal
AW	Adam Woolfit
BW	Bryan Wharton, London
DW	David H. Wells
D'LW	D'Lynn Waldron
EW	Eric L. Wheater
JW	John Wright
LW	Leslie Woodhead
MW	Margaret Watkins
PW	Patrick Ward
RW	Rod Williams
TW	Tony Ward, St. Albans
IY	Ian Yeomans
JY	John Yates
RY	Richard Young
LZ	L. Zatecky

PICTURE AGENCIES/SOURCES

A	Alpha, London
AP	Associated Press, London
APL	Aspect Picture Library, London
AS	Allsport, Morden, Surrey
BBC HPL	BBC Hulton Picture Library, London
BBC RT	BBC Radio Times, London
BC	Bruce Coleman Ltd., Uxbridge, Middx
CH	Camerapix Hutchison, London
CP	Camera Press, London
CS	Colorsport, London
DT	Daily Telegraph, London
F	Fotofolio (Rex Features), London
FI	Foto International, London
FSP	Frank Spooner Pictures, London
IB	Image Bank, London
JM	Joseph Mulholland, Glasgow
LC	Library of Congress, Washington
LE	London Express and News Service
M	Magnum Archive, Paris
MC	Mansell Collection, London

MEPL	Mary Evans Picture Library, London
N	Network, London
NFA	National Film Archive, London
NPG	National Portrait Gallery, London
P	Popperfoto, London
PA	Press Association, London
PP	Palace Pictures, London
PS	Photo Source, London
R	Rapho, Paris
REP	Reading Evening Post
RF	Rex Features, London
S	Sotheby's, London
SAPL	San Antonio Public Library
ST	Sunday Times Colour Magazine, London
SGA	Susan Griggs Agency, London
SF	Scope Features, London
SI	Syndication International, London
SPL	Science Photo Library, London
T	Transworld Features
UBN	UPI/Bettman Newsphotos
Z	Zefa, London

PICTURE LIST

Page numbers in bold type

Endpapers Beach exercises (EE) **M 2**
Nuba courtship dance, Sudan (LR) **3**
Boxer Marvin Hagler (PP) FSP U.S.
Olympic swimmer (GB) FSP **4** First
women's body–building world
champion, Lisa Lyón (RM) **5** Daly
Thomson long–jumping FSP **6**
Bushman hunter, Namibia (BS) A **7**
Returning from the hunt, Ethopia CH
Australian Aborigines around fire CH
8 Dinka mother washes baby son (SE)
CH Father holding child on one hand
(HCB) M Old woman and young girl,
Spain (GQ) **M 10** Artist's models,
London (DM) ST **12** Tall man, Paris
(PMd) R **13** Dwarf wedding
(Wells–Skyline) RF Mr and Mrs Dustin
Hoffman (J–PD/DA) Rizzoli Press **14**
Little ballerinas IB **15** Lady Tyger
Trimiar, world champion boxer (TD)
AS Czech athlete Jarmila
Kratochvilova AS **16** Identical twins
married to identical twins (RG) **17**
Parents and infant (LZ) Z **18** Dancers in
prenuptial rite portraying sun and
moon, Papua, New Guinea **M 19**
Japanese military parade (KK) FSP

THE HAIR **20** Long–haired women,
Massillou Museum, Ohio **21** Theda
Bara in Sin, 1915 F **22** Multi–coloured
man, Los Angeles FSP Fantasy hairdo
(PT) IB **23** Girl with Afro hair (DAk)
SGA Waist–length blond hair (NB) IB
Dancer exercising (JC) IB **24** British
Comedians, Morecombe and Wise
(ST) REP **25** Lionel the Lion Man MC
Chinese acrobats SAPL **26** Children, S.
Africa CH Chinese child CH **27**
Nigerian girl (JR) CH Boy in Welsh
graveyard M White–bearded man FSP
28 Boy pulling girl's hair PS
Anti–Vietnam protester seized by hair
(BU) **M 29** A Nazi–collaborator reviled
in the streets of Chartres after the
liberation of France in 1944 **M 30** Bald
old man (BM) IB Boy George with long
hair (MO) RF Boy George with short
hair (HSs) **31** African woman using
beer cans as hair rollers (MFl) IB Punk
with spiked hairstyle (BG) FSP Punk
having his hair dried (RSG) **32** English
footballer, Bobby Charlton on
massage table (RG) 'Four ways to

disguise a bald head' (Artist Peter Till) BBC RT **33** English Town Clerk adjusting wig SGA Japanese actor donning wig CH **34** Duncan Goodhew and girlfriend RF Olympic champion swimmer Duncan Goodhew – whose bald head is not shaved RF **35** American footballer (TD) AS Yul Brynner RF Telly Savalas RF Buddhist monk (FG) CH **36** Man scratching back of head (BL) N British prime minister Margaret Thatcher RF

THE BROW **37** Monk praying IB Chimpanzee (DB) BC **38** Thai girl (EW) IB **39** British actor Robert Morley (RG) British television presenter Bruce Forsythe RF Girl holding her tummy (DR) IB Spanish peasant with wrinkled brow (GQ) M **40** Marc Chagall (PH) Collection Albright–Knox Art Gallery Buffalo, New York, Gift of Seymour H. Knox 1976 F **41** Soccer Player Ian Bowyer receiving jabs (RG) **42** Shilluk King with beadlike brow mutilation, Sudan CH African with scored forehead CH African woman with ornamented brow (PC) A Dinka initiation ceremony, Sudan (SE) CH **44** Cocked eyebrow (RJ) **45** Brow Shrug, Via del Corso, Rome (RK) M **46** Jonathan Routh (GH) DT Jean Harlow NFA **47** Brooke Shields (T) RF Brooke Shields dressed as a man RF Werewolf from 'Company of Wolves' PP **48** Old woman with hand to brow (LF) R Indian infant in head compressor c.1860 MEPL

THE EYES **49** Jacky Coogan NFA ('The Kid' Charles Chaplin Estate) **50** Traditional mask of gold and pearls, Saudi Arabia A **51** Black Howler monkey (RW) BC Girl in muslin mask (DWH) IB Girl with flushed Apak girl CH **52** Chinese boy (dA) M **53** Boy in tears (WB) M **54** Brown eye Z Blue eye Z **55** Baby crawling (EA) M David Bowie RF **56** Woman in hospice, Ivry, France (MF) M **57** Salvadore Dali FSP Pope John–Paul II (ReL) (c) SPADEM **58** Miners, Tynewydd, Wales (JW) A Barry Humphries as Edna Everidge RF **59** Elton John in red cap RF Girl in red helmet (RSf) IB Girl with multi-coloured glasses (IM) IB Ceremonial headdress, Napende initiation, Zaire (SE) CH **60** Couple gazing at each other (GP) M Hostile stares (JK) M **61** Prostitute 'protected' by eye mobile (MN) SGA **62** Indian dancer, Shanta Dhananjayan (PF) SGA Steve McQueen with narrowed eyes RF Benny Hill and Corinne Russell RF **63** Marty Feldman in habit RF Girl winking (WK) IB **64** Punk eye make-up (SRG) Turk with blackened eyelids (drawing from Bulwer, 1650) Author's Collection

THE NOSE **65** U.S. Senator Sam Rayburn and friend AP **66** Boxer Marvin Hagler FSP Chinese man PS Australian Aborigine (HS) FSP French Academicians (MF) M **67** Man in red turban from Rajasthan (JPP) FSP **68** Steve McQueen and Ali McGraw RF Sophia Loren, unable to resist touching a child's nose (FV) FSP **69** Picasso and his son Claude (RC) M Italian father and son (RK) M **70** Caricature of Yvette Guilbert by Leandre MEPL Girl with rosette in her hair (HL) IB Mother and baby (SLa) SGA **71** Michael Jackson, as he was RF Michael Jackson as he is now, after a

nose–bob RF **72** Cartoon: 'The Nose' (Artist Peter Till) BBC RT Screen version of Cyrano de Bergerac NFA (Universal Pictures) Le Palace discothèque, Paris (GQ) M **73** Marlon Brando NFA **74** Nose–rub between nurse and patient, Russia (F) FSP Indian woman with nose rings (JL) A Ethiopian woman with nose decoration CH **75** The Princess of Wales in New Zealand (TG) Nambicuara boy with nose pins (JP) CH Hell's angel with nose ring (DS) M **76** Children reacting to teargas, Londonderry (GP) M Child wrinkling her nose (Chim) M

THE EARS **77** The Great Reclining Buddha, Rangoon (BB) M **78** Veteran of Verdun (MF) M Ear nibble (AP) **79** Leonard Nimoy as Dr. Spock in 'Star Trek' FI General de Gaulle FSP Back–lit long hair (US) IB **80** A loud engine, France (HCB) M **81** John Paul Getty II (Wheeler–Sipa) RF **82** Kenyan woman's ear with split lobe (PT) IB Mali woman with huge earrings CH Punk with multipierced ear (RSG) Bororo woman with multipierced ears (JW) CH **83** William Shakespeare, by unknown artist NPG **84** President Reagan doing Donkey Ears UBN Richard Ingrams at Nigel Dempster's ear (AG)

THE CHEEKS **85** Oboe player, Chad CH Fashion model Jerry Hall and daughter (H) RF **86** Rosy–cheeked baby (AW) SGA Girls and flowers, Peru (LDG) IB Girl with suntanned cheeks (GL) IB **87** Rosy-cheeked boy (GP) M John Wayne FI **88** Egg head (SD) Marilyn Monroe (EE) M **89** Frank Sinatra kissed by fan, 1944 P **90** Sir Matt Busby and Bobby Charlton celebrating football victory (RG) Frenchmen greeting with a kiss (RK) M **91** Yap woman with painted cheek CH Nigerian boy with scarred cheek (HS) FSP Dervish boy with skewered cheeks CH **92** J.K. Galbraith at UNESCO conference (dA) M Congratulatory cheek pinch, Portugal (JG) M

THE MOUTH **93** Cleo Rocas (AD) A Vietnamese lady (MF) M Elderly American woman (HCB) M **94** Chimpanzee expressions (MK) C (c) Time Life Inc. The accusation, Germany 1945 (HCB) M **95** Queen Elizabeth II and President Reagan (Walker) FSP Irish Republican youth shouting abuse (PM) M General de Gaulle FSP **96** Three mouth expressions (EH) M Gurner gurning SI **97** Clowns, Seville (JK) M **98** Lips under hat, Haiti (MFL) R Mick Jagger (ALz) C Yellow curls and red lips (SDe) IB **99** A bite on the cherry (O) IB Jewelled tooth (MN) SGA Akho lady with blackened teeth (RL) CH **100** Peasant woman (GvB) SGA **101** Denture cleaning, Portugal (SS) M **102** Twiggy on high stool (BGl) M Child painting (SS) M Van Halen (LH) FSP Devotee of Kali, India (MM) CH Mi–Careme carnival night at Le Palace discotheque, Paris (GQ) M **103** Open–air Mass (RC) M Couple at Henley Regatta (RoL) SGA **104** Yawning priest, Poland (JSv) CP **105** Jack Nicholson with cigar T Beer–spitting competitor (JGl) REP **106** The language of kisses MEPL Pope John Paul II kissing the ground (IB) M Mother feeding her baby by mouth (AV) IB **107** Ethiopian woman with

plates worn in earlobes and lips (LW) Granada/CH Lip plate in place (LW) Granada/CH Arab woman in yashmak, Abu Dhabi (RA) SGA **108** Mouth Shrug (C) R

THE BEARD **109** Man with long beard MEPL Ethiopian aiming a spear CH **110** Bearded American folk singers RF De Brazza's monkey (RW) BC Spot–nosed monkey (J/DB) BC Three English noblemen: Sir Charles Napier (top), First Baron Paget (middle), Sir Thomas Roe (bottom) NPG **111** Bearded Arab CH Melpa tribesman with painted face (MM) CH Bearded Chinese man (RSG) Sparse–bearded black (RSG) **112** Amsterdam barber (ES) M Bearded lady – Miss Annie Jones–Elliot, the 'Esau Lady', b.1865 in Virginia; exhibited in public from the age of two **113** Sophia Loren and Cary Grant (EH) M Row of male chins (HCB) M **114** Errol Flynn NFA Salvador Dali (PH) M Hitler in 1923 BBC HPL Olympic spectator (FD) FSP **115** Cowboy with side whiskers (D'LW) IB Groucho Marx NFA ('Duck Soup' Paramount Pictures) Sikh's coiled moustache, Rajasthan (RK) M Italian farmer (HCB) M James Cagney's chin chuck NFA ('Yankee Doodle Dandy' United Artists)

THE NECK **117** Woman's neck (GBo) IB **118** Nude profile (JS) **119** Alfred Hitchcock, 1962 (PH) M Nouveau ballet de Maurice Bejart (Versele) FSP **120** Female neckwear in Germany c. 1630 MEPL Geisha neck decoration, Japan M **121** Paul Getty and friend P **122** Giraffe woman, Burma (BB) M **123** Elaborate neck decoration, Kenya (HS) FSP **124** Talking round a pillar (RK) M **125** Craning necks in art gallery (HCB) M **126** Heads down in defeat (RG) Geisha girl bowing (MM) CH **127** Necking couple (RMc) SGA Pygmy woman (JV) FSP **128** Neck clasp (JK) M Father and son wrestling (JG) M

THE SHOULDERS **129** Marilyn Monroe NFA Boxer Frank Bruno (S) **130** Jerry Hall modelling for Dior RF Nude girl on chair (J-LA) FSP Muscleman Keijo Reiman, Finland AS **131** Darren Nelson of the Vikings AS Smiling girl (IT) IB Comedian Kenny Everett RF **132** Dress for beach or casino, 1899 MEPL Grace Jones, painting by Jean Paul Goude, Quartet Books. **133** Old lady and peacock (AW) SGA Couple in a crowd (LF) M **134** Victor and runner–up hugging after an Olympic final AS Couple on Rio beach (RoL) SGA Dinka couple CH Frank Sinatra and wife (Tr) RF **135** Cilla Black hugging herself RF Anna Ford and Mark Boxer (AD) A **136** Laughing woman (D) LC

THE ARMS **137** Children on ladder, Sarawak (TH) M **138** Ballerina (AC) IB **139** Girl with globe (JS) Body–building champion Arnold Schwarzenegger (GeB) M Schwarznegger and ballerina (GeB) M **140** Suzanne Dando (TW) **141** Charlton Heston NFA ('Ben Hur' MGM) Girl with shaved armpit IB Morris dancers (HSs) **142** Nuba ritual fighting, Sudan (LR) **143** Australian fast bowler, Jeff Thompson (RG) Olympic athlete, Tessa Sanderson (TD) AS **144** Pope John Paul II (BB) M Holy man by the Ganges (JL) APL Olympic swimmer, Michael Gross (TD) AS

THE HANDS **145** Pianist Arthur Rubenstein's hand (EH) M **146** Adult and child hands (Petit Format/Nestle) SGA **147** X–ray of hand SPL Pair of hands, dye–stained (HU) IB Couple holding hands (GC) IB **148** Yehudi Menuhin's hands (S) Helen Keller and President Eisenhower UBN **149** Muhammed Ali's fist (TH) M **150** Audrey Hepburn (CB) S Amal guard, Beirut (AM) FSP **151** Presidents Reagan and Geymayel (Markel) FSP English Rugby player (RG) **152** Palmist's crib sheet MC

THE INDEX